studies in physical and theoretical chemistry 49

METALLIC SUPERLATTICES
Artificially Structured Materials

studies in physical and theoretical chemistry 49

METALLIC SUPERLATTICES

Artificially Structured Materials

Edited by

TERUYA SHINJO
TOSHIO TAKADA

Institute for Chemical Research,
Kyoto University, Uji,
Kyoto-fu 611, Japan

ELSEVIER
Amsterdam — Oxford — New York — Tokyo 1987

ELSEVIER SCIENCE PUBLISHERS B.V.
Sara Burgerhartstraat 25
P.O. Box 211, 1000 AE Amsterdam, The Netherlands

Distributors for the United States and Canada:

ELSEVIER SCIENCE PUBLISHING COMPANY INC.
52, Vanderbilt Avenue
New York, NY 10017, U.S.A.

Library of Congress Cataloging-in-Publication Data

Metallic superlattices.

    (Studies in physical and theoretical chemistry ; 49)
    Includes index.
    1. Layer structure (Solids) 2. Superlattices as
materials. I. Shinjo, Teruya, 1938-   . II. Takada,
Toshio, 1922-   . III. Series: Studies in physical
and theoretical chemistry ; v. 49.
QD921.M4695 1987   530.4'1   87-20113
ISBN 0-444-42863-1 (U.S.)

ISBN 0-444-42863-1 (Vol. 49)
ISBN 0-444-41699-4 (Series)

Printed in The Netherlands

studies in physical and theoretical chemistry

Other titles in this series

Preface

The recent progress in evaporating techniques for preparing ultrathin films has made it possible to control each layer thickness on an atomic scale. By alternate deposition of two or more elements, artificial superstructures can be fabricated. The term "superstructure" is well known as a periodicity in a crystal with a wavelength longer than the unit cell dimension and various examples have been found in natural crystals. Although the term has been in use for a long time, it sounds rather current since artificially structured multilayers have attracted attention as a new class of superlattice. The subject of this book is artificial superlattices formed from metallic elements. Hereafter the term superlattice is used to refer to all multilayers with short wavelength compositional modulation, regardless of whether they are epitaxial or not.

We are interested in artificial superlattices mainly as a model system for fundamental research and as a new material which cannot exist in nature. By controlling the structure on a monolayer scale, we are able to design the sample to fit the study of a specific problem in solid state physics. In the following chapters, many examples verify that metallic superlattices are highly attractive in basic studies.

The advances made in semiconductor superlattices are described in a recently published comprehensive textbook, "Synthetic Modulated Structures", edited by Chang and Giessen (1985, Academic Press). In comparison, studies on metallic superlattices have just taken off but are developing very rapidly. From this standpoint, it may be rather premature to issue a textbook on metallic superlattices. It is certain that many new superlattices are already under intensive study and many more publications will appear in the following years. Nevertheless, we believe the publication of this book is worthwhile in providing the reader with an overview of metallic superlattices. Namely, one can see in what combination an artificial superstructure can be constructed, to which extent the structure can be tailored, and what are the physical and chemical properties. At the same time, it is also recognized that a vast area remains unexplored. This book is intended as an orientation not only for people in physics but also for those engaged in materials research in chemistry and technology. We hope this book will serve as a guide for new materials production and believe various successful applications in technology will appear in the near future.

This book is organized into seven chapters and an appendix. Chapter 1 is an overview of metallic superlattice studies, which serves as an introduction in general. Sample preparation and characterization techniques, a review of physical and chemical properties, and some perspectives on metallic superlattices are presented. In Chapter 2, structure analyses by means of x-ray diffraction are described. The neutron diffraction technique, introduced in Chapter 3, is especially useful for studying the magnetic properties. The application of nuclear magnetic measurements, NMR and Mössbauer spectroscopy, to the study of superlattices is described in Chapters 4 and 5. Chapter 6 deals with the superconducting properties of superlattices. Theoretical works are surveyed in the last chapter. In the Appendix, papers published up to 1986 are listed and also indexed according to the combination of the superlattice constituents.

We are grateful to Drs. N. Hosoito and N. Nakayama for their assistance in preparing this book. We also thank Mrs. F. Mizusaki for grammatical corrections and Miss K. Yamamoto for word-processing of the manuscripts.

Teruya SHINJO
and
Toshio TAKADA

List of Contributors

Malcolm R. BEASLEY, *Department of Applied Physics, Stanford University, Stanford, California 94305, USA*

Yasuo ENDOH, *Department of Physics, Faculty of Science, Tohoku University, Sendai 980, Japan*

Yasuhiko FUJII, *Faculty of Engineering Science, Osaka University, Toyonaka, Osaka 560, Japan*

Charles F. MAJKRZAK, *Brookhaven National Laboratory, Upton, New York 11973, USA*

Vladimir MATIJASEVIC, *Department of Physics, Stanford University, Stanford, California 94305, USA*

Teruya SHINJO, *Institute for Chemical Research, Kyoto University, Uji, Kyoto-fu 611, Japan*

Kiyoyuki TERAKURA, *Institute for Solid State Physics, University of Tokyo, Roppongi, Minatoku, Tokyo 106, Japan*

Hiroshi YASUOKA, *Institute for Solid State Physics, University of Tokyo, Roppongi, Minatoku, Tokyo 106, Japan*

Contents

Chapter 1

Overview of Metallic Superlattice Studies

T. SHINJO
Kyoto University

1.1 INTRODUCTION

Multilayered films are usually prepared by alternately depositing two elements using the vacuum deposition or sputtering techniques. A compositional modulation along the film normal is constructed if the following conditions are satisfied.

(1) Each layer thickness is controlled on an atomic scale.

(2) The deposited material forms a layered structure.

(3) Interdiffusion or interlayer chemical reaction is sufficiently suppressed.

In the past, x-ray opticists produced multilayers by combining heavy and light elements, W and C for instance, and tried to fabricate periodic structures with wavelengths not available in the lattice spacings of natural crystals. However, they measured the characteristics only as x-ray mirrors and were not concerned with the physical properties. In general, the vacuum condition before the 1960s was not high enough to produce reliable samples. On the other hand, semiconductor superlattice studies were initiated around 1970 by Esaki's group(1) and high vacuum techniques have since been introduced in the field of material fabrication. The advanced vacuum deposition method of preparing single crystal superlattices is called molecular beam epitaxy (MBE). Recent papers report that even a monolayer-monolayer superlattice of GaAs/AlAs has been successfully synthesized(2). Epitaxial and atomically flat layers are produced by slow deposition under ultrahigh vacuum (UHV) conditions.

While the history of metallic superlattices is rather long, it is only recently that significant developments in this research have been made. As is shown in Fig. 1.1.1, the number of publications on metallic superlattices has increased remarkably in these last five years. A list of these papers appears in the Appendix of this book. Among the fundamental physical properties of metallic superlattices, magnetism and superconductivity have been studied

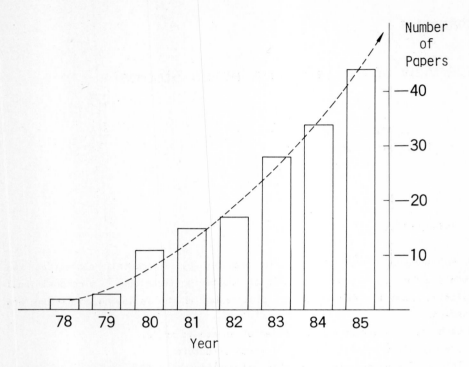

Fig. 1.1.1 Output of metallic superlattice studies per year.

extensively. As will be discussed later, artificial superlattices are useful
samples to investigate magnetic anomalies at interfaces and properties of two-
dimensional magnetic systems. Very recently, single crystal superlattices have
been synthesized from rare earth elements and the dependence of magnetic
structure on the artificial compositional modulation is being studied. In the
field of superconductivity, a commensurating effect to the superstructure is
observed in the temperature dependence of the upper critical field. It is
expected that the lattice dynamical properties of superlattices are more or less
unusual. Actually some reports suggest anomalously high elasticity. Chemical
properties are also of great interest. In fact, the initial purpose of
preparing metallic superlattices was to study thermal diffusion.

Figure 1.1.2 shows the combinations of metallic elements. The combination
used for making a superlattice and the number of publications reporting on that
particular combination are encircled. References classified by combination are
listed in the Appendix at the end of this book. The types of binary phase
diagram are indicated by A, B, C, and D. In category A, two elements form solid
solution in wide compositional ranges. In the case of B, mutual solubility is

Fig. 1.1.2   Combinations of metallic elements used for making superlattices. The numbers indicate how many papers were published by 1985.   A, B, C, and D refer to the type of binary phase diagram; rare earth metals are not included in this chart.

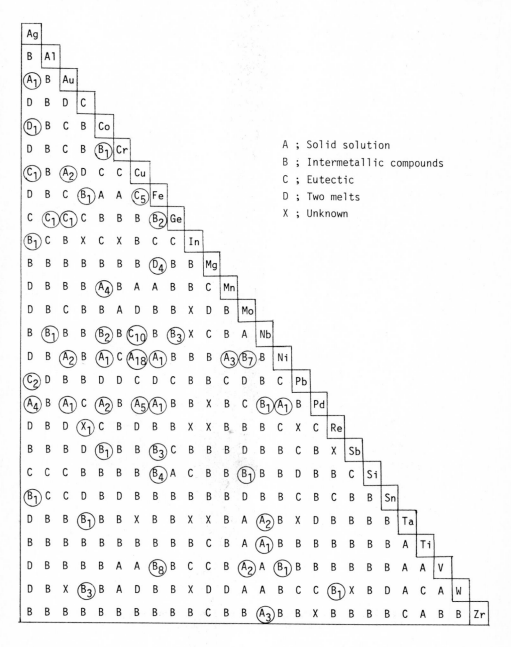

A ; Solid solution
B ; Intermetallic compounds
C ; Eutectic
D ; Two melts
X ; Unknown

4

limited but some intermetallic compounds exist.   Category C is for eutectic
systems; homogeneous mixing is only possible in the liquid state.   Category D is
for two-melt systems, which cannot be homogeneous even in the liquid state.   The
notation X is used to show that there has not been any report on the phase
diagram.   Perhaps most of the X combinations belong to category D.   Here we
notice that the formation of artificial superlattices is possible not only in
the A group but also in other categories. Already there are a number of circles
in the chart but intensive studies have been carried out for only a few
combinations, such as Ni/Cu, Fe/V, and Cu/Nb.   Rare earth elements are not
included in the chart, although some reports were published in 1985 and a number
of publications on superlattices from rare earth metals are appearing in 1986.
Besides these there are some other elements which may be used to prepare thin
films.   Certainly there are a great number of unexplored combinations.   If one
intends to prepare multilayers consisting of three elements, the number of
possible systems is nearly infinite.

From a structural viewpoint, artificial superlattices are classified into
two groups.   One is the epitaxial type and the other, non-epitaxial (Fig.
1.1.3).   The category A combinations can form epitaxial stacking rather easily
and the growth of a single crystal superlattice is possible.   In the other
categories, for the most part, the epitaxial relation cannot be held and each
layer may have a different structure.   Even when each layer is amorphous, the
superstructure can be established.

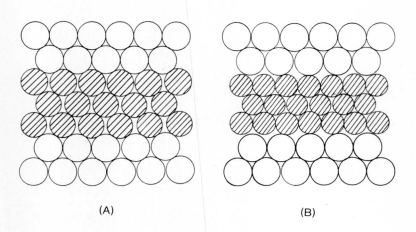

(A)   (B)

Fig. 1.1.3   Schematic illustration of cross-sections
of (A) epitaxial and (B) non-epitaxial superlattices

As for syntheses of new materials, artificial superlattices are very promising. Usually materials prepared through chemical procedures are in conditions near thermal equilibrium. Apparently, superlattices are far from the equilibrium. Non-epitaxial types are especially thermally unstable. In general, non-equilibrium states have not yet been studied adequately and we expect to find an interesting new material among them. Well-known examples of the non-equilibrium phase are amorphous and supersaturated phases produced by rapid quenching or ion-implantation. In contrast to these materials with random structures, we can microscopically design the structure of superlattices and control the properties. Thus, novel materials not only for fundamental research but also for technological use can be produced by the technique of multilayer deposition.

1.2 SAMPLE PREPARATION

Simply stated, the higher the vacuum, the better the quality of the obtained sample. Primarily the grade of the vacuum system to be used is determined by how much money can be spent. Significant technical difficulties are encountered only when we try to attain a region higher than 10^{-11}Torr. Even in a rather low vacuum, metallic films may be produced if the deposition rate is very fast. However, UHV is necessary for superlattice preparation when we want to control each layer thickness on an atomic scale.

Figure 1.2.1 is a schematic illustration of the sample preparation chamber being used at Kyoto University(3). In UHV, metal atoms ejected from the hearth have no collision object and fly in straight lines, forming an atomic (or molecular) beam. The thickness of each layer is therefore controlled by a shutter, which chops up the atomic beam of evaporated metallic elements. The shutter is driven rapidly by a pneumatic force, but the deposition rate cannot be too fast if we want to prepare monolayers or if an accuracy of 1Å is required for reproducibility. The usual deposition rate is between 0.1 and 0.5Å sec^{-1}; the deposition of one monolayer takes several seconds or more. The deposition should be slow enough to form a flat surface and to prevent interdiffusion. But, on the other hand, the surface will be contaminated by the adsorption of residual gas. Therefore UHV is necessary. Usually, the heating for evaporation makes the vacuum condition worse than the so-called base vacuum. Practically speaking, the region of 10^{-9}Torr would be a satisfactory condition during the deposition. A high evacuation speed in the 10^{-9}Torr region is preferable to an extremely high base vacuum. In this sense, a cryopump is useful.

To evaporate metallic elements with high melting points, electron beam (EB) guns are adopted instead of Knudsen cells, which are often used in MBE systems for making semiconductor superlattices. Since EB guns directly heat the evaporation source, no reaction takes place between the source and the water-cooled hearth. Compared to Joule–heating crucible methods, EB guns have another advantage in that the volume of the heated part is very limited and therefore the vacuum does not deteriorate much. Liquid nitrogen shrouds covering the heated places are also very helpful in maintaining a good vacuum.

In the system in Fig. 1.2.1, there are two reciprocal shutters on each hearth, which move independently. A quartz oscillating thickness sensor located near the substrate measures each layer thickness and a programmable thickness monitor controls the shutter motions according to a program such as the following; material A is deposited for a$\overset{\circ}{A}$, followed by an intermission of 5 sec.

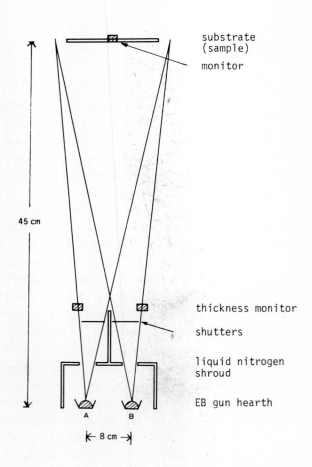

Fig. 1.2.1 Main part of the UHV deposition system.

then material B is deposited for b$\overset{\circ}{A}$, and the process is repeated. Thus each layer thickness is accurately controlled. Hereafter the structure of a superlattice is expressed by the nominal thickness, $[A(a\overset{\circ}{A})/B(b\overset{\circ}{A})]_n$, where n means the number of bilayers. Commercial thickness monitors using quartz sensors have a resolution limit of 1$\overset{\circ}{A}$. However, the reliability of measurements on absolute thickness values has been considered fairly uncertain. Experimental results on multilayered films, on the other hand, have shown that the accuracy of thickness measurements is rather good. For example, Fig. 1.3.2 in the next section compares the wavelengths of the compositional modulation designed by a thickness monitor with those established in synthesized multilayers which were estimated from x-ray diffraction. The good agreement between the designed and observed values proves the reliability of the thickness estimation.

The temperature of the substrate is very crucial. The optimum temperature depends on the components of the superlattice. In order to prevent intermixing, the temperature should not be too high. In fact, Fe/Mg superlattices can be prepared when the substrate temperature is lower than room temperature(4). If the substrate is continuously rotated, inhomogeneity of the sample due to geometrical conditions is excluded. However, the rotation of a substrate with cooling is technically difficult. On the other hand, for preparing single-crystal-like superlattices, the substrate temperature must be kept high enough for crystal growth. The Bell group prepared Gd/Y superlattices in the following manner: a buffer Nb layer was grown epitaxially on a sapphire substrate at a rather high temperature (700°C) and then Gd and Y were alternately deposited repeatedly on a cool substrate (220°C)(5). This procedure seems to be the prototype for preparing a single crystal superlattice.

In addition to vacuum deposition, sputtering is another well-established technique for preparing thin films and has been especially utilized in recent years to produce amorphous films. From a target, atoms are sputtered out by ionized argon gas. Fig. 1.2.2 shows a setup being used at Tohoku University(6). Two or more sputtering guns are equipped and substrates are placed on a rotating table. Deposition rates for sputtering are normally faster than evaporation and can be kept reasonably constant. By maintaining a constant rotating speed, a multilayered film with a periodic structure can be prepared. The wavelength of the artificial periodic structure can be easily shortened by making the rotation speed faster.

In comparison with vacuum deposition, sputtered atoms are supposed to have higher energies and therefore intermixing with the substrate seems to be more probable. According to Schuller et al., however, the energy of incoming atoms can be drastically reduced if the distance between the target and substrate is sufficiently large(7). Actually they prepared Nb/Cu superlattices with very short wavelengths, around 20$\overset{\circ}{A}$ at the minimum(8). Sputtering is useful for producing amorphous layers and in fact Kazama et al. prepared several amorphous/amorphous superlattices(6). In the case of sputtering, alloys or compounds can be used as the target, which is one of the merits of this technique. Thus, it is possible to prepare alloy/alloy superlattices. There are several demerits, however. For one thing, it is technically difficult to control the substrate temperature. Furthermore, it is impossible to completely exclude the inclusion of argon gas. Even though the amount is very small, the included gas impurities may play a crucial role in stabilizing amorphous structures.

Chemical vapor deposition (CVD) is another technique to synthesize superlattices. Its usefulness has already been recognized for semiconductor films. In some cases, this technique can produce an atomically flat monolayer with good crystallinity. However, the condition for crystal growth in this method is much nearer to thermal equilibrium. It seems rather difficult to avoid intermixing. So far, no successful result has been reported with metallic systems and the applicability of this method will probably not be very wide.

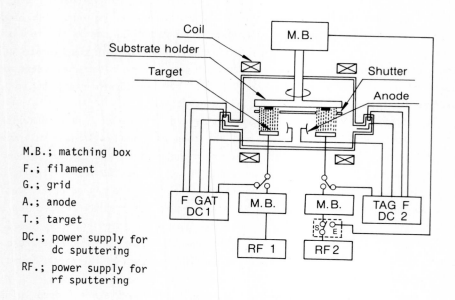

M.B.; matching box
F.; filament
G.; grid
A.; anode
T.; target
DC.; power supply for
    dc sputtering
RF.; power supply for
    rf sputtering

Fig. 1.2.2 *Dual source sputtering system, consisting of two components of dc-triod and/or rf-planar magnetron sputterings(6).*

1.3 CHARACTERIZATION

X-ray diffraction is a simple method to check whether or not an artificial superstructure exists in an alternately deposited film. If the lattice spacing is d, diffraction peaks will be observed at angles θ satisfying the well-known Bragg equation, $2d\sin\theta = n\lambda$, where λ means the x-ray wavelength. Since the period of artificial compositional modulation is usually much longer than the lattice spacings in natural crystals, the peaks are observed in the small angle region. X-ray diffraction patterns for some Fe/Mg superlattice samples are shown in Fig. 1.3.1(9). The patterns clearly indicate that artificial

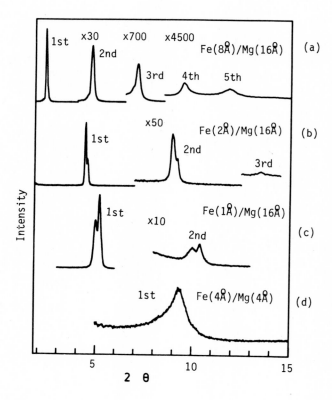

Fig. 1.3.1 X-ray diffraction patterns in the small angle region for some Fe/Mg superlattices(9). The design values (nominal layer thicknesses) are (a) Fe(8Å)/Mg(24Å), (b) Fe(2Å)/Mg(16Å), (c) Fe(1Å)/Mg(16Å), and (d) Fe(4Å)/Mg(4Å).

superstructures have been built from the mutually insoluble combination, Fe and Mg. The wavelength of compositional modulation in the prepared sample is estimated from peak angles larger than 2°, so that the refraction effect may be neglected. In Fig. 1.3.2, the observed values are compared with the design values of the bilayer period determined by the thickness monitor. The agreement is excellent. For such metal elements, the thickness estimation is reliable because the sticking coefficients for both the substrate and the thickness sensor are nearly 100%. For the calculation of thickness, the densities of deposited layers have been assumed to be the same as the bulk values. This result suggests that the assumption is reasonably good.

In order to obtain information on the crystal structure of individual layers, peaks in the middle angle region have to be studied. In the case of epitaxial superlattices, sharp peaks with satellites are observed. On the other hand, non-epitaxial ones show rather broad peaks and amorphous superlattices none. For the crystallographic characterization, the sharpness, rocking curve, and relative intensities of x-ray diffraction peaks, including higher order ones

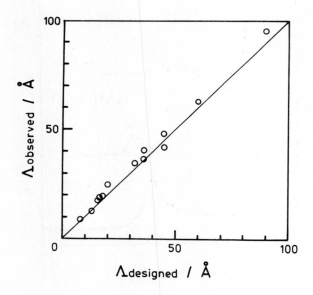

Fig. 1.3.2   Comparison of designed and observed wavelengths in Fe/Mg superlattices.   Observed values are estimated from small angle   x-ray diffraction peaks(9).

are analyzed. The use of intense beams of x-rays from synchrotrons for the structural characterization of superlattices will soon become more popular. By selecting the optimum wavelength x-ray, the contrast in the diffraction pattern may be enhanced. The extended x-ray absorption fine structure (EXAFS) technique gives information on the local environments and is especially useful for samples with highly disordered structures(10) or those whose individual layer thicknesses are in the monolayer region. Details of x-ray characterization are presented in Chapter 2.

Cross-sectional observation by transmission electron microscope (TEM) is useful for confirming visually the stripe patterns established in a superlattice sample. Figure 1.3.3. shows a cross-section of an Fe(4Å)/Mg(16Å) sample. Although each Fe layer is only two atom layers thick, it is fairly flat and continuous. The homogeneity of the sample, from the first to the last layer, can be confirmed. (Only a portion of the enlarged photograph is shown in the figure.) The undulation of the pattern might give the impression that the interfaces are not atomically flat. However, because each stripe is a perspective of several hundred atoms, each plane can be regarded as extremely flat. The TEM sample is made by cutting the superlattice sample with a microtome. The structure before cutting should be more beautiful than that shown in the picture.

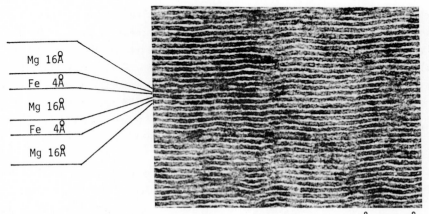

Fig. 1.3.3 TEM observation of a cross-section of an Fe(4Å)/Mg(16Å)
superlattice, taken by A. Ishizaki, Canon Inc.(9).

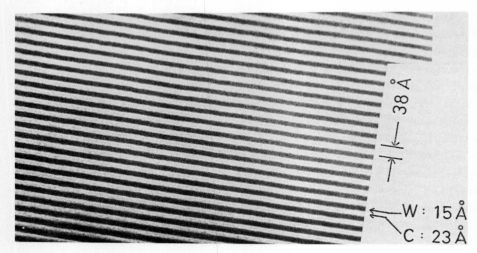

Fig. 1.3.4   TEM observation of a cross-section of a W/C superlattice(11).

Another example shown in Fig. 1.3.4. is a cross-section of a W/C superlattice, which was prepared by sputtering by the NTT group(11). Even amorphous materials can be used for an artificial superlattice with very short wavelength. In this case, the nominal bilayer thickness is 38Å, which is twice as thick as the Fe/Mg superlattice in Fig. 1.3.3. The stripe pattern is very clear and each line is apparently straight. The flatness of each interface is visually confirmed. Such a cross-sectional observation by TEM is very useful for evaluating a superlattice sample as long as it can be cut very thinly. For epitaxial superlattices, one can study the lattice coherency at the interface from the lattice image observation by high resolution TEM measurements. At the present time, however, it is so difficult to cut the sample satisfactorily thin and to treat it properly that this technique is not yet a routine one for characterizing superlattices.

Rutherford backscattering is a rather common technique in thin film characterization. By analyzing the energy of reflected He^+ particles, the species of the target element can be identified and the depth from the surface estimated from the energy loss. However, the depth resolution limit is around 20Å, which is not enough to study a compositional modulation with a very short wavelength. This technique is useful for wavelengths longer than 100Å, where x-ray diffraction becomes inconvenient. It is also useful for detecting certain impurities such as Ar, which may be included in the films prepared by sputtering. Using this technique, Kazama et al. found a certain amount of Ar entrapped in Si layers (see Fig. 1.3.5)(12).

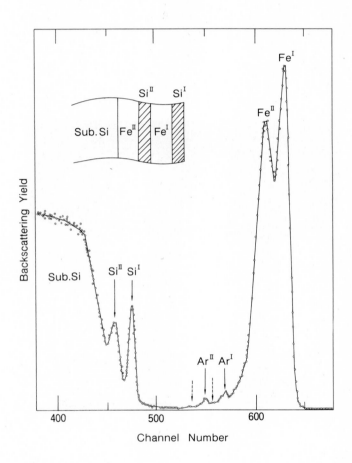

Fig. 1.3.5 Backscattering spectrum of 3 MeV He$^+$ ions from a film with four layers of Fe^{I,II} and Si^{I,II}. The solid arrows indicate the Ar signals from the deposited Si layers and the dashed arrows indicate the expected positions of Ar signals from Fe layers(12).

14

Artificial superstructure can also be checked by Auger electron microscopy. If the surface is continuously etched by Ar gas sputtering, the intensity of the Auger signal should show an oscillation. However, the depth resolution cannot be more than 20Å and a quantitative analysis seems to be very difficult. If the surface is etched smoothly, the stripe pattern is observed in the scanning Auger micrograph. An example is shown in Fig. 1.3.6, which is the result of a Cu/Ti superlattice observed by Hatano et al.(13). Although only qualitative, the homogeneity of the sample's structure is visually confirmed. If the surface has been successfully etched with a very slow gradient, the stripe pattern with a periodicity of some 100Å can be seen by the naked eye.

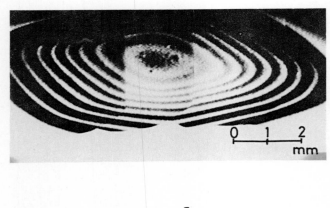

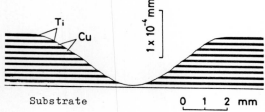

Fig. 1.3.6 Scanning Auger micrograph of a Cu/Ti superlattice(13);
Λ = 66Å.

The quality of a superlattice sample should be evaluated not only from a crystallographic but also from a chemical viewpoint. A crystallographically perfect superlattice has a single-crystal-like structure with a unique wavelength. On the other hand, chemical perfectness requires a compositional modulation of a square wave mode with 100% amplitude. These two characteristics are independent of each other and quite often we are faced with the dilemma of improving the crystallinity and at the same time preventing intermixing. Kwo et. al. estimated that the interdiffusion at the interface of Gd/Y superlattices was confined to two atom layers(5). According to McWhan, this range of mixing is near the minimum for interfaces in actual superlattices, and even in single crystal semiconductor superlattices, similar mixings take place(14).

If the range of interdiffusion is limited to a few atomic layers, it is hard to estimate the chemical profile at the interface with a structural analytical method. For this purpose, a microscopic magnetic study using hyperfine field measurements is very useful. Applications of Mössbauer spectroscopy and nuclear magnetic resonance are described in Chapters 4 and 5.

1.4 PHYSICAL PROPERTIES

Multilayers prepared by alternate deposition of magnetic and non-magnetic materials are suitable specimens to study the magnetic behaviors of interfaces and also two-dimensional systems. The fundamental problems of interface magnetism are as follows: (i) the magnitude of the local magnetic moment at the interface site atom, (ii) the temperature dependence of the interface magnetization, and (iii) the contribution of the interface to the magnetic anisotropy. If the magnetic layer can be reduced to monolayer thickness, a two-dimensional magnetic system will be obtained. The behavior of magnetic spins on a two-dimensional lattice has been an attractive subject in basic physics and many theoretical predictions have been published. So far experimental studies have been carried out on natural crystals with layered structures but a problem is interlayer interaction, which is normally not negligible. On the other hand, interlayer interaction in artificial superlattices is a controllable parameter. If the non-magnetic spacing layers are thick enough, interlayer interaction can be regarded as zero.

As in the case of x-ray diffraction, neutron diffraction can be used to examine the periodic structures of superlattices. For magnetic/non-magnetic superlattices, distribution of the magnetic density is estimated and thus magnetic anomaly at the interface is detected. Neutron diffraction from superlattices is a new tool for studying magnetism at interfaces or in monolayers. Also, hyperfine field measurements by Mössbauer spectroscopy and nuclear magnetic resonance provide specific information for the study of magnetism. Generally speaking, however, a sample which contains only one interface layer, or one monolayer, is not convenient to obtain reliable information. From hyperfine field measurements on superlattices having several hundred layers, we can obtain fruitful information on the microscopic magnetic properties with a good signal-to-noise ratio. In an ideal case, where the chemical modulation at each interface is abrupt, the magnetic behaviors of the interface can be studied from the results of such measurements. In actual superlattices, more or less chemical mixing probably occurs at the interfaces. Before discussing the intrinsic interface effects, we should clarify the compositional profile at the interface. The usual experimental methods for structural analysis or macroscopic magnetic measurements do not have enough resolution to detect very limited mixing, such as intermixing only within a few atomic layers or a compound layer of monolayer thickness. A microscopic magnetic study with hyperfine field measurements is very useful for detecting mixing in a limited range and for speculating on the local atomic arrangements. The details of superlattice studies by neutron diffraction, Mössbauer spectroscopy, and nuclear magnetic resonance are described in succeeding chapters. In this chapter only a brief survey is presented.

An example of a magnetic/non-magnetic superlattice is Ni/Cu, which has been studied rather extensively by several groups in the United States. From their results with ferromagnetic resonance measurement, Thaler and Ketterson concluded that the magnetic moment of Ni atoms at the interface sites is bigger than the standard bulk value(15). This led to an increased interest in metallic superlattices in the field of magnetism. Gyorgy et al. investigated the magnetism of Ni/Cu superlattices in detail and eventually refuted the enhancement of the magnetic moment(16). Ni and Cu are similar elements with the same fcc structure and can form solid solutions in the entire compositional range. As a result of epitaxial growth, the superlattices also have fcc structures. Since the samples were prepared on substrates at rather high temperatures, crystal growth was ensured but interdiffusion could not be inhibited. Inevitably the chemical profile became sinusoidal instead of the ideal square wave mode. It was impossible to obtain a monolayer of Ni with ferromagnetic behaviors.

Succeedingly, several combinations among 3d metals have been studied, such as Mn/Ni(17), Mn/Co(17), Cu/Fe(18), and Co/Cr(19). Fe/V is an example of a bcc/bcc epitaxial superlattice(20). Since a cold substrate was used, intermixing at each interface was fairly limited. Hyperfine field measurements suggest the mixing is within three atom layers. However, the size of the crystallites is very small, on the order of 100Å. In the case of epitaxial superlattice fabrication, we are faced very often with the dilemma of ensuring crystal growth without intermixing. Artificial compositional modulation is established even when the nominal Fe layer thickness is reduced to one monolayer, i.e., Fe(2Å)/V(16Å), but certainly the structure of the "monolayer" is not perfect. Mössbauer results showed that Fe atoms in the Fe(2Å) layers are non-magnetic, having lost the local magnetic moment. However, we cannot compare these results with theoretical predictions unless the structure of the monolayers is atomically flat, chemically pure, and the area is infinitely large. In any case, the actual Fe monolayers in the Fe/V system are non-magnetic and therefore not appropriate for the study of magnetic properties of a two-dimensional system.

The situation of a superlattice from an insoluble combination, Fe/Mg, is different. The magnetic interface effect is very small; that is, Fe atoms at the interface layer contacting with Mg have magnetic moments similar to the bulk value. Fe monolayers sandwiched in-between Mg layers are evidently ferromagnetic and the local magnetic moments are not much reduced. Therefore this system can be used to study the magnetism of a two-dimensional lattice. Generally speaking, chemically sharper interfaces are obtained in a non-epitaxial superlattice than in an epitaxial one. However, the abruptness of the compositional change at the interface is not a priori certified even in non-epitaxial cases. Although Fe and Mg are immiscible in the thermal equilibrium state, it has been reported that amorphous phases were prepared in a certain compositional range(21). Mössbauer spectroscopic studies are very useful for elucidating the chemical profile at the interface, as described in Chapter 4.

Peculiar magnetic characteristics have been found in Fe monolayers sandwiched in-between Mg layers. The easy direction is along the surface normal instead of the in-plane usually observed in magnetic films. The magnetization is stable only at rather low temperatures; magnetic hyperfine splittings in the Fe(2Å) and Fe(1Å) samples collapse at about 55K and 35K, respectively. However, the bulk susceptibility at room temperature is significantly large, suggesting short-range spin correlations. The obscurity of the magnetic transition may be considered one of the characteristics of low dimensional magnetic systems. X-ray diffracion studies suggest that the structure of the Fe monolayer is different from that of Mg but the details are not clear yet. Speculation of the interface structure is usually very difficult in non-epitaxial superlattices. In order to further develop an analysis, a knowledge of the crystal structure is essential.

Another way to prepare a ferromagnetic monolayer is with the Mn/Sb system(22). These elements are not ferromagnetic but the intermetallic compound MnSb is known to be strongly ferromagnetic. When the two elements are simply alternately deposited without any heat treatment, this compound has been found to be formed at the interface. X-ray analysis shows a distinctive periodic structure even when the Mn layer thickness is 1Å, which corresponds to the amount of a MnSb monolayer. The sample including MnSb monolayers is prepared by depositing Mn(1Å) and Sb(60Å) alternately (Fig. 1.4.1). Since the structure of the MnSb monolayer is epitaxial to the Sb layer, it can be determined. The monolayer sample is ferromagnetic at 4.2K and very interestingly the easy direction is perpendicular to the film plane (see the upper graph in Fig. 1.4.1).   In both of the above cases of ferromagnetic monolayers, the spontaneous magnetization is along the normal to the film, which is probably the result of surface (or interface) anisotropy,   originally   proposed   by   Néel(23). Magnetization in such monolayers is stable only at low temperatures.   If the

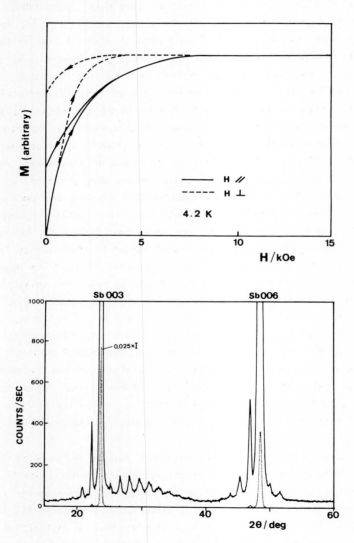

Fig. 1.4.1   X-ray diffraction pattern of [Mn(1Å)/Sb(60Å)] × 110. Satellites around sharp Sb peaks are due to the artificial periodic structure with a wavelength of 60Å.   The upper graph   shows the magnetization vs. external field curves at 4.2K, applied parallel and perpendicular to the film plane(22).

thickness is greater than two atomic layers, the direction of the magnetization turns to the in-plane.   Recently ferromagnetic films with a perpendicular easy direction have attracted interest from a technological standpoint, but these monolayers have too low Curie temperatures.   Carcia et al., working with Co/Pd superlattices, have found that perpendicular magnetization appears if the thickness is less than 8Å and that Co(8Å) films are ferromagnetic at 300K(24). This suggests the possibility of using metallic superlattices for magnetic recording media.

Mn and Sb are mutually insoluble but can form intermetallic compounds (MnSb and Mn_2Sb).   This system, therefore, belongs to category B in Fig. 1.1.2.   Fe/Sb and Co/Sb superlattices have also been prepared, and their phase diagrams are more or less similar to that of Mn/Sb.   In these superlattices of category B, intermetallic compounds are easily formed at the interfaces.   The reactivity between Fe and Sb or Co and Sb is much weaker than between Mn and Sb.   However, according to the results from Mössbauer and NMR measurements, approximately one monolayer of Fe (or Co) at the interface has an electronic structure very similar to that of an intermetallic compound.   Nominal monolayers of Fe or Co sandwiched in-between Sb can be prepared but are found to lose the magnetic moments(25).   In other words, a model system for the study of two-dimensional magnetism can be found in Mn/Sb but cannot be realized in Fe/Sb or Co/Sb superlattices.

So far most of the studies on the magnetic properties of 3d metal superlattices have been concerned more with the structure of thin layers than with the coherent effects caused by the artificial periodic structures. Theoretically,   interesting predictions have been published.   For instance, Lambin and Herman studied the magnetism of a Ni_3Fe/FeMn superlattice, which is a combination of ferromagnetic and antiferromagnetic layers(26).   However, it is extremely difficult to prepare a sample of sufficient quality that can be compared with such theoretical studies.   On the other hand, magnetic structures of rare earth metals are known to have long-range periodic structures, which must be greatly modified in superlattices if the layer thickness is to be comparable to or less than the intrinsic magnetic wavelength.   Recent reports suggest incommensurate magnetic structures in Dy/Y superlattices(27) and long-range antiferromagnetic coupling through Y layers in Gd/Y superlattices(28). Such rare earth superlattices are interesting subjects for neutron diffraction and also for magnetic x-ray diffraction using synchrotron radiation.   The usefulness of neutron diffraction for the study of magnetic properties in superlattices is described in more detail in Chapter 3.

Superconductivity is another basic physical property of metallic systems. Compared with magnetism, the coherence length is much longer and dimensional effects already appear in films with a thickness of several 100Å.   Since the coherence length is temperature dependent, that is, decreasing with a decrease in temperature, the same superlattice film exhibits a two-dimensional behavior at low temperatures and a three-dimensional one at higher temperatures. Such dimensional crossover is manifest in the H_{c2} vs T curves of the systems, Nb/Ge(29), Nb/Cu(30), and V/Ag(31). A commensurability effect is pointed out in the V/Ag studies. Superconductivity studies on artificial superlattices and also very thin films are described in detail in Chapter 6 and it is shown that superlattices are useful specimens to investigate basic parameters in superconductivity physics, such as coherence length, penetration depth, proximity effect, and localization. Since superconductivity is a very sensitive property which depends on the quality or homogeneity of the sample, it is useful for the characterization of the superlattice sample. For instance, the sharpness of the superconducting transition is a reliable indication of sample homogeneity. Figure 1.4.2 shows the results of resistive and inductive measurements for a $[V(40\text{Å})/Ag(20\text{Å})]_{30}$ sample(32). The transition width in the resistivity is about 20mK. Below the transition temperature, perfect diamagnetism is confirmed in the real component (χ') of the susceptibility when a magnetic field is applied perpendicular to the film plane. The imaginary component (χ'') forms a sharp peak around the transition point. If the sample is inhomogeneous, the peak χ'' becomes broad and thus the quality of the sample is checked from the sharpness of peak χ''.

We hope to find new kinds of superconducting materials in metallic superlattices since there are a great number of unexplored combinations as shown in Fig. 1.1.2. Unusual new materials may be produced by combining metals and non-metals. However, among superlattice samples so far prepared, no remarkable success has yet been achieved in elevating the transition temperature.

Since the early days of metallic superlattice studies, attention has been focused on the elastic properties. A considerable enhancement of the biaxial modulus of Cu/Ni superlattices was reported in 1977(33). Although other measurements followed to confirm this result, some criticism has also been raised. Since superlattice samples are always very thin films, a quantitative estimation of elastic constants is technically difficult. The experimental results are not regarded as conclusive even though it is plausible for superlattices to have unusual elastic properties because each interface can store significant stresses. On the other hand, Schuller et al. found lattice softening in Ni/Mo, Nb/Cu, and V/Ni superlattices by Brillouin scattering measurements and also in V/Ni superlattices by surface acoustic wave measurement(34). They claimed that the softening is a general phenomenon in such bcc/fcc superlattices and the enhancement may occur only in epitaxial systems.

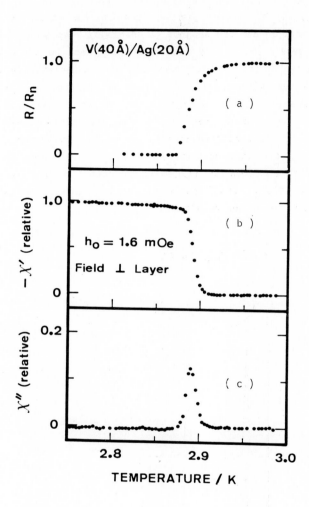

Fig. 1.4.2 Superconductive transition of $[V(40\text{Å})/Ag(20\text{Å})]_{30}$; (a) temperature dependence of resistance (R_n: resistance at 4.2K) and (b) temperature dependence of (b) real and (c) imaginary parts of ac susceptibility with a magnetic field perpendicular to the film plane(32).

1.5 CHEMICAL PROPERTIES

Since artificially structured superlattices are not in the thermal equilibrium state, the structures are dependent on the substrate temperature during sample preparation and are subject to change by heat treatments. In the preparation, the temperature of the substrate should not be too high. Figure 1.5.1 shows the x-ray diffraction patterns of Fe/Mg superlattices prepared on substrates at different temperatures. The sample prepared on a substrate at below room temperature shows distinct small angle peaks due to the superstructure, but there is no peak at all when the substrate temperature is 100°C. The effect of the substrate temperature on the superstructure has not yet been systematically studied since a quantitative comparison of peak intensities in different samples is generally rather difficult. However, regarding the intensity of small angle diffraction peaks, a low substrate temperature often gives a better result.

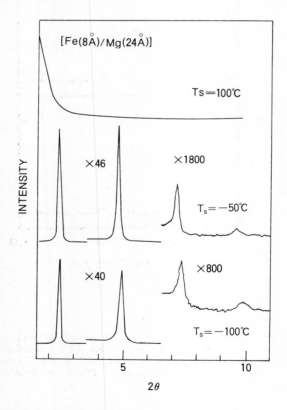

Fig. 1.5.1 X-ray diffraction patterns in the small angle region for [Fe(8Å)/Mg(24Å)] samples. The substrate temperatures during sample preparation are shown in the figure(4).

On the other hand, if formation of a single crystal superlattice is intended, the substrate temperature should be raised to some extent. Durbin et al. succeeded in preparing a Nb/Ta single crystal superlattice film on a substrate heated to 700 ∿ 900°C, though how much interdiffusion had taken place at the interface was not examined(35). Semiconductor superlattices, GaAs/AlAs for instance, are prepared on substrates at about 700°C but interdiffusion is considered to be less than a few atomic layers(14).

Empirically speaking, artificial periodic structures can be established in almost all combinations of metallic elements if the substrate temperature is low enough and the individual layer thickness is designed to be not less than 30Å. The critical value of the wavelength to hold a distinct compositional modulation depends on the constituents and was found to be around 20Å in several cases. The wavelength of compositional modulation can be shortened to 10Å only in limited cases. Otherwise films with amorphous-like structures are obtained. By setting the wavelength at a sub-monolayer thickness, we can intentionally prepare a homogeneous amorphous film.

If the free energy of the artificial superlattice state is higher than that of the amorphous state, the superlattice turns out to be amorphous when treated with heat at a moderate temperature. This was investigated by Johnson et al., who reported preparing several amorphous combinations, such as Zr/Co, starting from multilayers(36). From the viewpoint of thermodynamics, the process of amorphous formation from a superlattice is interesting and is considered a new method for preparing an amorphous material. However, it is not clear what kind of difference there is between an amorphous film prepared by this method and a simultaneously deposited one.

The effect of heat treatment after sample preparation cannot be completely explained by drawing an analogy between this and the effect of substrate temperature. The situation of epitaxial superlattices consisting of mutually soluble elements is rather simple. Interdiffusion is accelerated at high temperature and each crystallite grows in size. Finally the sample becomes homogeneous and the modulated structure disappears. The effect of thermal annealing on Ni/Cu superlattices was considered simply a decrease in the amplitude of chemical modulation. If the combination of elements belongs to category B in Fig. 1.1.2, where formation of intermetallic compounds is probable, compound formation takes place at the interface. Examples of category B are Fe/Sb, Co/Sb, and Mn/Sb, whose chemical profiles are discussed in Chapters 4 and 5.

In the case of non-epitaxial superlattices, belonging to category C or D, the artificial superstructure is, of course, decomposed into two independent phases by heavy heat treatment, but heat treatment at a moderate temperature may improve the crystallinity of each layer while keeping the periodic structure.

24

For instance, Fe/Mg superlattices are fairly stable at 100°C, though the superstructure is not formed if the substrate temperature is 100°C. It has sometimes been found that thermal annealing at a moderate temperature enhances the intensity of the small angle x-ray diffraction peaks. An example is shown in Fig. 1.5.2, where the diffraction peaks in the small angle region after heating at 155°C for 52hrs are much stronger than as-prepared(37). The peaks in the middle angle region suggest crystal growth in the individual layers to a certain extent. However, the artificially structured periodicity has not decomposed but is somewhat refined. In order to utilize a superlattice (e.g., W/C films) as an x-ray mirror, the reflectivity of the small angle x-ray beam should be as high as possible. For this purpose, perhaps the best result is obtained when heat treatment is applied to a supperlattice prepared on a rather cold substrate.

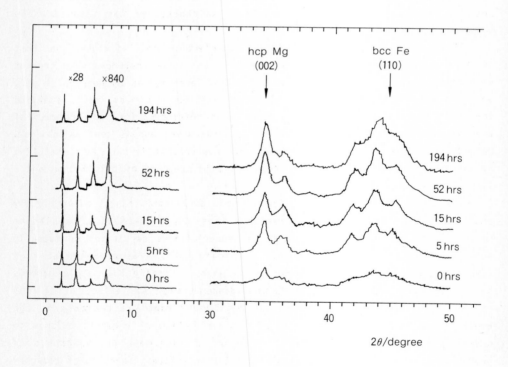

Fig. 1.5.2 Thermal annealing effect on the X-ray diffraction pattern for [Fe(15Å)/Mg(30Å)] × 100. After heating at 155°C for the time length written in the figure, the X-ray diffraction was measured at room temperature.

Absorption of hydrogen in Nb/Pd(38) and Nb/Ta superlattices(39) has been studied. Hydrogen loading was accomplished in situ using an x-ray furnace and the induced strain modulation was measured by x-ray diffraction. The result suggests that the solubility of hydrogen into superlattices is much larger than that into bulk metals under the same condition. Hence an interesting chemical property of superlattices has been revealed. Although metallic superlattices are attractive subjects for studying chemical properties, this research has been limited thus far to a few studies on hydrogen loading and thermal annealing. As yet there has been no report concerned with the catalytic ability of metallic superlattices, although this problem seems to be very interesting from a technological standpoint.

1.6 PERSPECTIVES

The empirical knowledge we have gained so far suggests that every combination of metallic elements bears some artificial periodic structures and has the potential of being created into a new material, as far as the substrate temperature during the sample preparation is not too high. The detailed structures of the superlattice, as already mentioned, depend greatly on the types of constituents, epitaxial, non-epitaxial, forming intermetallic compounds, and so on. If the number of metallic elements suitable for thin film preparation is around 50, the possible binary combinations for producing superlattices turn out to be 50 X 50/2. The combinations shown in Fig. 1.1.2 are only a part of this total. The number of possible ternary systems, i.e., superlattices composed of three elements, is much higher. Not only metallic elements but also non-metallic substances can be joined to make exotic new materials. Already some examples of superlattices consisting of metal/oxide(40) or oxide/oxide layers(41) have been reported. Indeed the unexplored area in the field of superlattices is extremely broad and therefore more effort should be made to examine superlattices of unknown combinations.

The growth process of the superlattice is far from the thermodynamical equilibrium state. Evaporated atoms are quenched at the surface of a cold substrate, which has an infinitely large mass. The problem is how does an incident atom dissipate the thermal energy and settle down at a certain site. If the thermal conductivity of the substrate is extremely good, the energy of the incident atom will dissipate immediately into the whole body of the substrate. Then the situation would be rather simple and the relaxation process of each single atom has to be considered. However, the real situation, especially in the initial stage of deposition, is much more complicated. The thermal energy of an incident atom is transferred to some atoms of the substrate (the other species) in the initial stage and also to previously deposited atoms (the same species). A certain amount of atoms may participate in the reconstruction of the surface but the size of such an assembly is hard to guess. It is very difficult to predict the product from thermodynamical considerations, although some attempts have been made by computer simulation of the deposition process(42). Further experimental investigations are required to examine the structures and properties of superlattices artificially prepared from unknown combinations.

On the other hand, the quality of superlattices needs to be improved. Generally, the vacuum condition during sample preparation should be kept as high as possible to avoid any serious problems caused by impurities. It is desirable to utilize in situ analytical techniques, such as Auger spectroscopy, refractive high energy electron diffraction (RHEED), and mass analysis.

In the case of epitaxial combinations, an ideal sample is a single crystal superlattice with chemically sharp and atomically flat interfaces. To obtain atomically flat surfaces, a technique that has been used in semiconductor superlattice preparation is worth referring to. If the intensity of the reflected RHEED beam intensity is measured during the growth of GaAs single crystal, an oscillation of the intensity corresponding to the growth of a monolayer is observed. GaAs type compounds are known to grow in a layer-by-layer fashion. The intensity of RHEED reaches a maximum when the surface atomic layer is perfect, or atomically flat with 100% occupation. On the other hand, if the surface atom layer is only about 50% occupied, the microscopic roughness of the surface is greatest and the RHEED intensity is minimum. Based on this principle, the intensity of specularly reflected RHEED was used to monitor the completeness of each deposited layer during the preparation of a GaAs(monolayer)/AlAs(monolayer) superlattice(2). As you can see, this technique may be used to improve the quality of metallic single crystal superlattices. Figure 1.6.1(A) shows the oscillatory feature of specularly reflected electron beam intensity during the growth of GaAs and (B), the intensity variation during the growth of the monolayer-monolayer superlattice. The variation of the Auger intensity and also the oscillation of electrical resistivity can be used to check whether or not the film is growing in a layer-by-layer fashion(43). However, using these parameters to control film growth in the case of superlattice preparation seems to be very difficult.

In what sense can we expect to obtain new materials from synthetic multilayers? As we near the end of this chapter, let us consider this question in general. If individual thicknesses are very large, e.g., 1,000Å, each layer should possess its original properties. A multilayer composed of material A (1,000Å) and material B (1,000Å) will show a sum of the properties belonging to both A and B. It is just A + B but cannot be anything more. Nevertheless, a

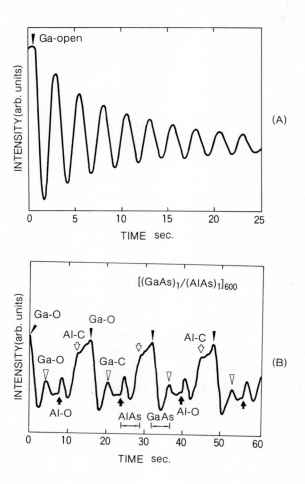

Fig. 1.6.1 (A) Intensity oscillation of the specularly reflected electron beam during the growth of GaAs. (B) Intensity variation of the specularly reflected electron beam during the monolayer/monolayer superlattice growth(2).

composite for technical use might be created if one combines different characteristics in such a way as to fit a certain purpose, for instance, combinations of ferromagnetism, superconductivity, or ferroelectricity.

If the individual layer thickness is in the region of some 100Å, the behavior of conduction electrons will be greatly affected. In the field of semiconductor superlattices, various attempts to modify the electronic structures have been successful. Quantum well structures have been constructed along the film normal, and two-dimensional electron gas systems, in the plane. On the other hand, in transport phenomena of metallic superlattices, it seems very difficult to observe an intrinsic quantum effect related to the periodic structure because synthetic metallic superlattices include a lot of crystallographic imperfections, such as grain boundaries, vacancies, and irregularities at the interface layers. Electrons are scattered very much by such imperfections and intrinsic effects due to the periodic structure are hidden. Meanwhile, multilayers can be used to study systematically the effects of localization.

The coherence length of usual superconducting materials is on the order of 100Å. Therefore, films with a two-dimensional character are easily prepared. For studies of dimensional crossover phenomena, crystallographic imperfections are not very serious. However, if the superconductor layer thickness is less than 100Å, T_c is reduced considerably because of the localization effect. An exotic superconductor is expected to be produced by combining two materials which play different but complementary roles to enhance T_c; one layer conducts superconducting current while the other works to create strong correlations between electrons. Superconducting superlattices with very short compositional wavelengths are interesting from another standpoint. They are regarded as single phases but may have unknown electronic band structures. Thus, a new superconductor should be found among them.

The wavelength of a superlattice to be utilized as a mirror for x-ray or neutron optics is on the order of 10Å. The recent progress in thin film fabrication techniques has renewed an interest in multilayers in this field. In addition, the very recent practical application of a strong synchrotron x-ray source has accelerated the development of x-ray experimental techniques and thus multilayers are required for many purposes. The wavelength of an artificial superlattice can be varied as desired and therefore can cover the region which is not available by natural crystals. Perhaps another merit is the freedom of choosing the elements constituting the superlattice. In the near future, a superlattice will not only be grown on an optically flat surface but also on a curved one with a special design. Already plans have been made to mount multilayers on a satellite to collect cosmic x-rays(44). For more details on the utilization of multilayers in x-ray optics, the references cited in the next chapter should be referred to.

By making use of epitaxy, one may obtain new materials since metastable phases are sometimes realized due to epitaxy. A well-known example of such a metastable phase is Fe layers epitaxially grown on a Cu substrate. As far as the Fe layer thickness is less than about 20Å, the structure is fcc, instead of the usual bcc, and the ferromagnetism has disappeared(45). Recent experiments by Brodsky are also instructive in this respect. He reported that a thin Cr layer ($\sim$20Å) sandwiched in-between Au layers takes an fcc structure(46). Normally Cr metal has a bcc structure and exhibits antiferromagnetic properties. In contrast, the epitaxial fcc Cr shows superconducting behavior whose T_C is about 4K. He also prepared Au/Pd/Au film and measured its properties, expecting the appearance of ferromagnetism or superconductivity caused by the expansion of the Pd lattice spacing. Actually an enhanced magnetic susceptibility was observed(47). These results suggest that one may obtain a new material by packing metastable layers in a multilayered structure. The properties are of great interest, particularly if the metastable phase has special characteristics such as superconductivity with high T_c, even though it is a minor fraction in the superlattice.

If the wavelength of an artificial periodic structure becomes very short, around 10Å or less, superlattices can have properties different from the original ones of the constituents. Then, the superlattices are regarded as ordered alloys of a new type. In the case of epitaxial superlattices, the local atom arrangements can be conjectured and the electronic structures can be theoretically studied. For instance, Jarlborg and Freeman calculated the band structure of Cu(3layers)/Ni(3layers) superlattices(48).(Fig. 1.6.2) As a matter of fact, at the present stage, the gap between an ideal structure postulated in the calculation and the actually prepared one is not very small. Nevertheless there is no doubt that computer physics will interplay more closely with materials research in this new field. The details of this kind of theoretical study are presented in Chapter 7. In the field of materials research, there is great interest in superlattices consisting of insoluble or eutectic combinations, since such elements cannot be homogeneously mixed on a microscopic scale. Most of these superlattices are of the non-epitaxial type and their energy states are far from thermal equilibrium. If the degree of non-equilibrium can be defined, one may say that non-epitaxial superlattices are of a higher degree than epitaxial ones. On the other hand, a demerit of non-epitaxial superlattices for the study of microscopic electronic structure is that local atomic arrangements cannot be easily estimated. As is described in the next chapter, the crystallinity of Fe/Mg superlattices is very poor when the individual Fe layer thickness is less than 10Å. It seems inevitable that the structure of non-epitaxial superlattices with extremely short periodicity

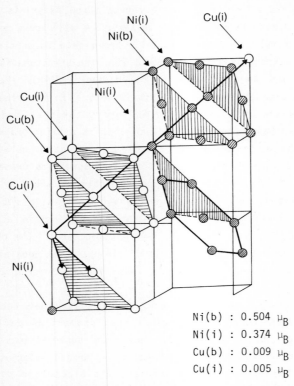

Ni(i)
Ni(b)
Cu(i)

Cu(i)
Ni(i)
Cu(b)

Cu(i)

Ni(i)

Ni(b)	:	0.504 μ_B
Ni(i)	:	0.374 μ_B
Cu(b)	:	0.009 μ_B
Cu(i)	:	0.005 μ_B

Fig. 1.6.2 Ni/Cu superlattice modulated along [III]. The calculated spin magnetization per atom at each site is shown in the figure. (b) and (i) indicate bulk and interface sites, respectively(48).

becomes somewhat amorphous.

In principle, superlattices are constructed not only from crystalline layers but also from amorphous ones. In fact, Kazama et al. prepared distinctive periodic structures from amorphous Fe(including C) layers and amorphous Si layers and showed that the wavelength can be shortened to about 15Å(8). Recently, Sato reported the fabrication of Tb/Fe superlattices with very short wavelengths(49). His result is very interesting from a technological standpoint since these films exhibit excellent magnetic properties suitable for magneto-optic recording media. It has been suggested that in the near future a possible application of superlattices in the technological field may be as a magnetic recording medium. In the meantime, Kerr effect studies on various superlattices are in progress(50).

So far we have prepared modulated structures with single wavelengths, which are the easiest to characterize. However, our technique for controlling the thickness on an atomic scale may be used to construct more complicated structures with sophisticated designs. In fact, very recently, the preparation of a quasiperiodic superlattice (the Fibonacci series) combining Mo and V has been attempted and the superconducting properties were reported(51). At least along the normal to the film, it is possible to prepare any type of compositional modulation. Moreover, in the near future, attempts will be made to construct modulated structures in three-dimension.

REFERENCES

(1) For semiconductor superlattice studies, see reviews in "Synthetic Modulated Structures" L. L. Chang and B. C. Giessen, eds., Academic Press 1985.
(2) H. Terauchi, S. Sekimoto, K. Kamigaki, H. Sakashita, N. Sano, H. Kato and M. Nakayama, J. Phys. Soc. Jpn. 54 (1985) 4576.
(3) T. Shinjo, N. Hosoito, K. Kawaguchi, T. Takada, Y. Endoh, Y. Ajiro and J. M. Friedt, J. Phys. Soc. Jpn. 52 (1983) 3154.
(4) T. Shinjo, K. Kawaguchi, R. Yamamoto, N. Hosoito and T. Takada, Thin Solid Films 125 (1985) 273.
(5) J. Kwo, E. M. Gyorgy, D. B. McWhan, M. Hong, F. J. DiSalvo, C. Vettier and J. E. Bower, Phys. Rev. Letters 55 (1985) 1042.
(6) N. S. Kazama and H. Fujimari, J. Magn. & Magn. Mater., 35 (1983) 86.
(7) K. Meyer, L. K. Schuller and C. M. Falco, J. Appl. Phys. 52 (1981) 5803.
     I. K. Shuller and C. M. Falco, "VLSI Electronics, Microstructure Science Vol. 4" (1982) 183.
(8) I. K. Shuller, Phys. Rev. Letters 16 (1980) 1597.
(9) K. Kawaguchi, R. Yamamoto. N. Hosoito, T. Shinjo and T. Takada, J. Phys. Soc. Jpn. 55 (1986) 2379.
(10) T. Claeson, J. B. Boyce, W. P. Lowe and T. H. Geballe, Phys. Rev. B29 (1984) 4969.
(11) K. Takei and Y. Maeda, Jpn. J. Appl. Phys. 24 (1985) 118.
(12) N. S. Kazama, H. Fujimari, S. Yamaguchi and Y. Fujino, J. Magn. & Magn. Mater. 35 (1983) 214.
(13) T. Hatano, K. Nakamura,Y. Asada,A. Oguchi and K. Ogawa, Jpn. J. Appl. Phys.
(14) D. B. McWhan, Ref. (1), p. 43.
(15) B. J. Thaler and J. B. Ketterson, Phys. Rev. Letters 41 (1978) 336.
(16) E. M. Gyorgy, D. B. McWhan, J. E. Dillon Jr., L. R. Walker and J. V. Waszczak, Phys. Rev. B25 (1982) 6739.
(17) M. B. Stearns, J. Appl. Phys. 53 (1982) 2436.
(18) H. M. Noort, F. J. A. den Broeder and H. J. G. Draaisma, J. Magn. & Magn. Mater. 51 (1985) 273.
     H. J. G. Draaisma, H. M. V. Noort and F. J. A. den Broeder, Thin Solid Films 126 (1985) 117.
(19) M. Takahashi, S. Ishio and Y. Notohara, Proc. 2nd Intern. Conf. Physics of magnetic materials, Jadwisin, 1984, World Sci Pub. Singapore (1985) p. 13.
(20) A structural study; Y. Endoh, K. Kawaguchi, N. Hosoito, T. Shinjo, T. Takada, Y. Fujii and T. Ohnishi, J. Phys. Soc. Jpn. 53 (1984) 3481.
     Hyperfine field studies on Fe/V superlattices are discribed in Chapters 4 and 5.
(21) A. M. van der Kraan and K. H. J. Buschow, Phys. Rev. B25 (1982) 3311.
(22) T. Shinjo, N. Nakayama, I. Moritani and Y. Endoh, J. Phys. Soc. Jpn. 55 (1986) 2512.
(23) L. Néel, Compt Rend. 237 (1953) 1468.
     A review on the studies of surface anistropy, for instance, is U. Gradmann, Appl. Phys. 3 (1977) 173.

32

(24) P. F. Carcia, A. D. Meinhaldt and A. Suna, Appl. Phys. Lett. 47 (1985) 178.
(25) Details for Fe/Sb and Co/Sb are described in Chapters 4 and 5.
(26) P. Lambin and F. Herman, Phys. Rev. B30 (1984) 6903.
(27) M. B. Salomon, S. Sinha, J. J. Rhyne, J. E. Cunningham, R. W. Erwin, J. Borchers and C. P. Flynn, Phys. Rev. Letters 56 (1986) 259.
(28) C. Vettier, D. B. McWhan, E. M. Gyorgy, J. Kwo and B. M. Buntschuh, Phys. Rev. Letters 56 (1986) 757.
   Results from neutron diffraction are described in Chapter 3.
(29) I. Banerjee, Q. S. Yang, C. M. Falco and I. K. Schuller, Phys. Rev. B28 (1983) 5037.
(30) S. T. Ruggiero, T. W. Barbee Jr. and M. R. Beasley, Phys. Rev. B26 (1982) 4894.
(31) K. Kanoda, H. Mazaki, T. Yamada, N. Hosoito and T. Shinjo, Phys. Rev. 33B (1986) 2052.
(32) N. Hosoito, Y. Yamada, K. Kanoda, H. Mazaki and T. Shinjo, Bull. Inst. Chem. Res. Kyoto Univ., 64 (1986) 235.
(33) W. M. C. Yang, T. Tsakalakos and J. E. Hilliard, J. Appl. Phys. 48 (1977) 879.
(34) R. Danner, R. P. Huebener, C. S. L. Chung, M. Grimsditch and I. K. Schuller, Phys. Rev. B33 (1986) 3696. Other papers on lattice dynamical properties are cited therein.
(35) S. M. Durbin, J. E. Cunningham, M. E. Mochel and C. P. Flynn, J. Phys. F. Metal Phys. 11 (1981) L223.
(36) M. Van Rossum, M. -A. Nicolet and W. L. Johnson, Phys. Rev. B29 (1984) 5498.
   R. B. Schwartz, K. L. Wong, W. L. Johnson and B. M. Clemens, J. Non-Cryst. Solids 61-62 (1984) 129. B. M. Clemens, ibid. 817.
   W. L. Johnson, M. Van Rossum, B. P. Dolgin and X. L. Yen, "Rapidly Quenched Metals", S. Steeb and H. Warlimont, eds., Elsevier Sci. Publishers 1985, p. 1515.
   K. Samwer, A. Regenbrechet and H. Schröder, ibid. p. 1577.
   H. Schröder, K. Samwer and U. Köster, Phys. Rev. Letters 54 (1985) 197.
(37) T. Ishihara, Y. Fujii, Y. Yamada, K. Kawaguchi, N. Nakayama and T. Shinjo, to be published.
(38) S. Moehlecke, C. F. Majkrzak and M. Strongin, Bull. Am. Phys. Soc. 28 (1983) 876, Phys. Rev. B31 (1985) 917.
(39) P. F. Miceli, H. Zabel and J. E. Cunningham, Phys. Rev. Letters 54 (1985) 917.
(40) M. H. Tanielian, J. R. Willhite and D. Niarchos, J. Appl. Phys. 56 (1984) 417.
(41) T. Terashima and Y. Bando, J. Appl. Phys. 56 (1984) 3445.
(42) R. Ramirez, A. Rahman and I. K. Schuller, Phys. Rev. B30 (1984) 6208.
(43) J. P. Renard, Nato Advanced Research Workshop on "Thin-film growth techniques for low dimensional structure" (1986) The proceedings is to be published by Plenum.
(44) Private communication from Prof. K. Yamashita, Osaka Univ.
(45) R. Halbauer and U. Gonser, J. Magn. & Magn. Mater. 35 (1983) 55.
(46) M. B. Brodsky, P. Marikar, R. J. Friddle, L. Singer and C. H. Sowers, Solid State Commun. 42 (1982) 675.
(47) M. B. Brodsky and A. J. Freeman, Phys. Rev. Letters 45 (1980) 133.
   M. B. Brodsky, J. Appl. Phys. 52 (1981) 1665 and Phys. Rev. B25 (1982) 6060.
(48) T. Jarlborg and A. J. Freeman, Phys. Rev. Letters 45 (1980) 653.
(49) N. Sato, J. Appl. Phys. 59 (1986) 2514.
(50) T. Katayama, H. Awano and Y. Nishihara, J. Phys. Soc. Jpn. 55 (1986) 2539.
(51) M. G. Karkut, J. -M. Triscone, D. Ariosa and Ø. Fischer, Phys. Rev. B34 (1986) 4390.

Chapter 2

X—Ray Diffraction Studies
on Metallic Superlattices

Y. FUJII
Osaka University

2.1   INTRODUCTION

When one starts an experiment on superlattices, it is always necessary to confirm the successful growth of the superlattice structure in advance.   X-ray diffraction is a very powerful and nondestructive method readily available   to see how regularly the alternately-deposited layers are stacked   and how sharply the interfaces are formed. On the other hand, metallic superlattices have been applied as novel optical devices in a wide   variety of fields, such as monochromators for synchrotron radiation x-rays   and mirrors for   solar x-ray telescopes. Their lattice-spacing and   scattering-power density can be tailored to match required specifications.   Thus x-rays are used for characterizing superlattices while   superlattices are used for handling x-rays.

X-rays scattered by individual constituent atoms give rise to interference depending upon the position of the scatterers. From a viewpoint of the structure thus probed by x-rays, we may classify the order inherent in the superlattice structure in "chemical" and "structural"   orders, as displayed schematically in Fig.2.1.1. The chemical order is a measure of the profile of the composition wave, i.e. amplitude and shape. The degree of order is defined to be high for a composition varying abruptly at the interface. On the other hand, a superlattice with a compositionally-graded interface has a low degree of order.   The structural order is further classified into three types. The first is an   "intralayer" order concerned with the periodicity of atomic ordering   within individual layers.   An amorphous layer has a low degree of order while it is high for a crystalline layer. The second is an "interlayer" order to evaluate how periodically individual layers are stacked in the growth direction.   The degree of order is higher for a structure with the smaller fluctuation of its superlattice periods throughout the film.   The   third is a "lateral" order, which is limited to crystalline layers.   This order becomes perfect for an epitaxially-grown superlattice being a single crystal as a whole. A superlattice consisting of many randomly-oriented grains in the lateral direction has a low degree of lateral order. A similar definition to characterize superlattices was previously made by several authors. McWhan (1) called the interlayer and lateral structural orders "coherence". It is defined as the coherence length in a superlattice where x-rays can be coherently diffracted without losing their phase relation. Barbee (2) made a more detailed classification necessary for evaluating the quality of a superlattice applied to figured x-ray optics. According to him, other quantities such as flatness, smoothness and cleanness of both substrate and deposited layers should also be considered.

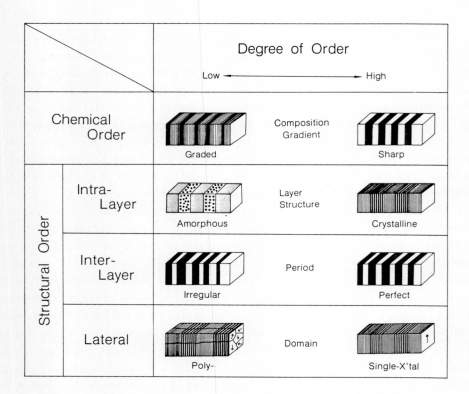

Fig. 2.1.1 Chemical and structural orders inherent in the superlattice structure. X-ray diffraction is a sensitive method to probe the degree of each order.

In this Chapter, we describe how to perform an x-ray diffraction experiment on a superlattice, how to deduce structural information from observed diffraction patterns and how superlattices are used as devices. The next section (2.2) deals with x-ray diffraction methods and techniques particularly useful for the study of superlattices. In Section 2.3, model calculations of diffraction patterns are made for the different degrees of the chemical and structural orders with reference to specific examples of superlattices. These calculations are based on the kinematical diffraction theory to show how diffraction patterns reflect the structural characteristics of superlattices. In Section 2.4, the dynamical diffraction theory is employed to calculate diffraction patterns for perfect or nearly perfect superlattices. Section 2.5 presents novel methods of synchrotron radiation x-ray diffraction recently advanced for studying and characterizing superlattices. A useful table for x-ray experiments is also attached at the end of this Chapter.

2.2 EXPERIMENTAL METHODS OF X-RAY DIFFRACTION

2.2.1 Reciprocal space

Let us briefly review diffraction conditions for x-rays. It is very convenient to understand diffraction phenomena in reciprocal space instead of real space. If a, b and c are taken as the unit lattice vectors specifying a unit cell in real space and they are assumed to be orthogonal for simplicity, the unit reciprocal lattice vectors are defined as

$$a^* = 2\pi(b \times c)/v, \quad b^* = 2\pi(c \times a)/v, \quad c^* = 2\pi(a \times b)/v, \tag{2.1}$$

where v is a unit cell volume, i.e. $v = a \cdot (b \times c)$. Any reciprocal lattice vector H is known to be expressed as

$$H = ha^* + kb^* + \ell c^* = 2\pi e/d_{hk\ell},$$

$$H = |H| = 2\pi/d_{hk\ell} = 2\pi[(h/a)^2 + (k/b)^2 + (\ell/c)^2]^{1/2}, \tag{2.2}$$

where $d_{hk\ell}$ is the lattice spacing of the $(hk\ell)$ plane and e is the unit vector normal to this plane. Now let us consider x-ray diffraction from a specimen as sketched in Fig.2.2.1. The incident x-ray beam with wavelength λ impinges upon a specimen and the direction of the beam changes from e_i to e_f ($|e_i| = |e_f| = 1$) after being diffracted elastically. The wave vectors of the incident and diffracted beams are given by $k_i = ke_i$ and $k_f = ke_f$ ($k = 2\pi/\lambda$), respectively.

The scattering vector is defined as

$$Q = k_f - k_i , \qquad Q = |Q| = 2k \sin\theta , \qquad (2.3)$$

where 2θ is the diffraction angle determined from the scattering triangle shown in Fig.2.2.1(b). Equations(2.2) and (2.3) lead to the well-known Bragg's law in kinematical diffraction, i.e. $2d_{hk\ell}\sin\theta = \lambda$ in real space, which is expressed as

$$Q = H , \qquad (2.4)$$

in reciprocal space. That is, Bragg reflection takes place when the specimen is placed in such a position that the scattering vector ends in any reciprocal lattice point.

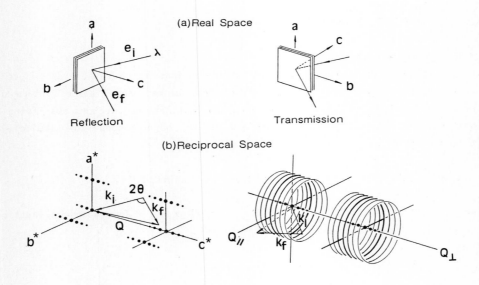

Fig. 2.2.1 *Diffraction geometry of x-rays with respect to a super-lattice expressed (a) in real space and (b) in reciprocal space. The reflection and transmission in (a) represent the configurations to probe the structures in the stacking and lateral directions, respectively. The left- and right-hand figures in (b) schematically show a single-crystal superlattice (epitaxially-grown) and a superlattice composed of many poly-crystals in the lateral direction, respectively.*

If the lattice constant of the **c** axis is chosen to be the superlattice period Λ along the layer-stacking direction, we have $c^* = 2\pi/\Lambda$ and the other two axes lying in the film plane. Figure 2.2.1(b) shows the scattering geometry for x-rays with respect to the reciprocal space of the superlattice. When the superlattice is composed of many poly-crystals in the lateral direction as shown in Fig.2.1.1, the reciprocal lattice points located on the principal c^* axis are still spots while other points spread out in powder rings as shown at right in Fig.2.2.1(b). Since in this case the reciprocal lattice is cylindrically symmetric with respect to the c^* axis, the reciprocal space can be expressed by two axes $Q_\perp$ and $Q_{//}$ which are taken along c^* and any direction in the film plane, respectively. In order to check the formation of the superlattice structure, a scan is first made along the principal $Q_\perp$ direction. When it is successfully grown, the first order Bragg reflection 001 due to the charge density contrast between two kinds of layers appears at the scattering vector $Q = \Lambda^* = 2\pi/\Lambda$, or at the scattering angle given by $2\theta = 2\sin^{-1}(Q_\perp/2k) = 2\sin^{-1}(\lambda/2\Lambda)$. Higher order reflections 00ℓ appear at positions expressed by integral multiples of Λ^* ($Q = \ell\Lambda^*$). One can easily understand the convenience of using Q instead of 2θ to express the peak position, because Q is a quantity inherent in the superlattice structure and independent of the wavelength of x-rays used.

2.2.2 Apparatus

A conventional x-ray diffractometer is schematically shown in Fig. 2.2.2(a). It is comprised of a sealed-off type x-ray tube (power 1∿2 kW) and a double-axis diffractometer equipped with a detector (GM, proportional or scintillation counter). The incident beam is either filtered with a metallic foil or monochromatized with a crystal. Figure 2.2.2(c) shows a double-axis diffractometer with a position-sensitive detector (PSD), which can collect diffraction patterns in a wide angular range (typically about 10 degrees of the diffraction angle) in a single measurement. This novel detector system is particularly useful for an in situ measurement of time-dependent phenomena such as the thermal annealing effect of a superlattice. Figure 2.2.2(b) is a more convenient diffractometer equipped with a Eurelian cradle to orient a specimen to any desired direction. Figure 2.2.2(d) is a so-called four-circle diffractometer installed on a rotating-anode x-ray source (power 18 kW). In addition, it is equipped with an analyzer crystal between the sample- and detector-positions to achieve a higher momentum (angular) resolution and reduce the background (3). These diffractometers are widely used in laboratories for evaluating the quality of prepared superlattices and for studying details of their structural characteristics. Recently much more brilliant x-ray sources

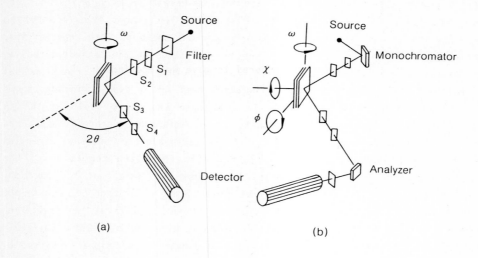

(a)

(b)

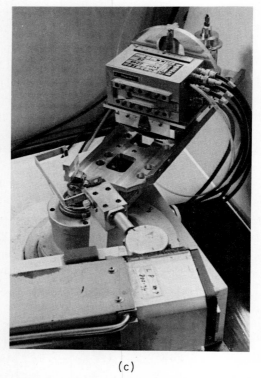

(c)

(d)

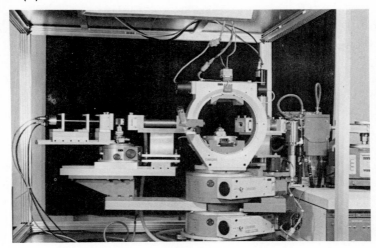

Fig. 2.2.2 Various x-ray diffractometers used for the study of superlattice structures. (a) A conventional diffractometer with two rotation axes ω (sample) and 2θ (detector). (b) An advanced four-circle diffractometer equipped with both a Eurelian cradle(φ,χ) and an analyzer crystal. Such a diffractometer installed on a rotating-anode x-ray source is shown in (d) [Institute for Chemical Research, Kyoto University]. (c) A double-axis diffractometer equipped with a position sensitive detector(PSD) which has a detection length 50mm and a positional resolution 0.2mm [Faculty of Engineering Science, Osaka University].

have become available as synchrotron radiation which provides a unique capability for studying superlattices as presented in 2.5.

2.2.3 Experimental remarks

(i) Selection of x-ray energy   The proper selection of x-ray energy to be used for a given superlattice is very important to obtain high quality data with a high signal-to-noise ratio. First of all, one must avoid excitation of fluorescence x-rays from constituent atoms by using an x-ray energy E lower than the absorption edge E_a (usually K- or L-edge). Here the conversion relation between the x-ray energy E and its wavelength λ or wave-number k is given by

$$E \text{ (keV)} = 12.398/\lambda(\text{Å}) = 1.9732 \, k(\text{Å}^{-1}). \tag{2.5}$$

For example, Cu-Kα radiation (E = 8.040 keV) is not suitable for a superlattice containing Fe with its K-edge at $E_K(Fe) = 7.111$ keV because the excited Fe-Kα fluorescence x-rays (E = 6.401 keV) produce a high background. In this case, the use of Fe-Kα radiation is recommended. The lower energy x-rays also shift the Bragg peak positions to higher angles so that the 1st-order peak is more easily measured.   In the Table   at the end of this Chapter, the energies of characteristic radiations and absorption edges are tabulated for various elements.   The second point to consider in selecting the x-ray energy is to have a higher contrast of scattering-power density between the two kinds of layers. This is   particularly important for a superlattice consisting of two elements which are near each other in the periodic table, such as Fe/Mn and Cu/Ni superlattices.   In such a case, the use of anomalous scattering is   very effective.   The atomic scattering factor f^0 is significantly modified near the absorption edge as $f = f^0 + \Delta f' + i\Delta f''$.   Figure 2.2.3(a) shows the wavelength dependence of $\Delta f'$ and $\Delta f''$ for Fe and Mn in the vicinity of the K-absorption edges [$E_K(Mn) = 6.538$ keV or $\lambda_K = 1.896$ A].   Figure 2.2.3(b) demonstrates how the 1st-order Bragg reflection of the Fe/Mn superlattice changes its intensity across the absorption edges.   This experiment was carried out by using the energy tunability of synchrotron radiation(4).   Details of this   work are presented in 2.5.1.

   (ii) <u>Correction factors</u>   In order to obtain the intrinsic intensity distribution, given by the square of the structure   factor $|F_H|^2$ versus reciprocal lattice vector H in the kinematical diffraction theory, the observed intensity I_{obs} must be corrected for various instrumental factors.   Usually one can write

$$I_{obs} = K\ A\ P\ L\ |F_H|^2 , \qquad\qquad\qquad (2.6)$$

scale factor   K = constant,

absorption factor

$$A = \begin{cases} \mu^{-1}[1 + \sin(\theta+\phi)/\sin(\theta-\phi)]^{-1} \\ \quad \times\ [1 - \exp\{-\mu t/\text{cosec}(\theta-\phi)-\mu t/\text{cosec}(\theta+\phi)\}] \\ \qquad\qquad\qquad\qquad \text{for reflection,} \\ \mu^{-1}[1 - \cos(\theta+\phi)/\cos(\theta-\phi)]^{-1} \\ \quad \times\ [\exp\{-\mu t/\cos(\theta-\phi)\} - \exp\{-\mu t/\cos(\theta+\phi)\}] \\ \qquad\qquad\qquad\qquad \text{for transmission,} \end{cases}$$

polarization factor

$$P = \begin{cases} (1 + \cos^2 2\theta)/2 & \text{for unpolarized beams,} \\ \\ (1 + \cos^2 2\theta_M \cos^2 2\theta)/2 & \text{for monochromatized beams,} \end{cases}$$

$$(2\theta_M : 2\theta \text{ angle of monochromator})$$

(a)

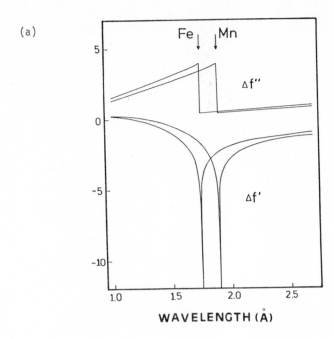

(b)

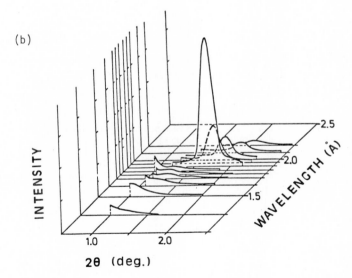

Fig. 2.2.3 (a) Wavelength dependence of the anomalous scattering factors $\Delta f'$ and $\Delta f''$ for Fe and Mn in the vicinity of the absorption edges $\lambda_K(Fe) = 1.743$ and $\lambda_K(Mn) = 1.896$ Å. (b) The 1st-order Bragg reflection observed on $[Fe(15Å)/Mn(50Å)]_{40}$ as a function of wavelength selected from continuous spectra of synchrotron radiation x-rays. [From Nakayama et al.(4)]

Lorentz factor

$$L = \begin{cases} 1/\sin 2\theta & \text{for single crystal,} \\ \\ 1/\sin^2\theta \; \cos\theta & \text{for powder crystal.} \end{cases}$$

In factor A, the angles θ and ϕ are defined as shown in Fig.2.2.4 and μ is a linear absorption coefficient averaged over the constituent atomic densities of the superlattice. In the transmission geometry, one must also take into account the absorption by the substrate.

In order to obtain the intrinsic peak profile, an instrumental resolution function must be deconvoluted from the observed peak. This process is crucial to evaluate the coherence length, i.e. how distantly stacked layers diffract x-rays coherently. As sketched in Fig.2.2.5, the longitudinal width $W_{//}$ is defined as the peak width in the direction through the origin, which is the so-called θ–2θ direction. On the other hand, the peak width perpendicular to the longitudinal one, measured by rocking the specimen, is called the transverse width $W_\perp$. As shown in the next section, $W_{//}$ is a measure of the coherence length while $W_\perp$ is observed as the sum of the mosaic spread of a superlattice and the peak-broadening due to the relevant coherence. These two contributions can be separated by using the difference of their Q-dependence. The former is proportional to Q while the latter is independent of Q. In Fig.2.2.5, the $W_{//}$'s along the $Q_\perp$ and $Q_{//}$ directions represent the coherence length for the interlayer and lateral orders defined in Fig.2.1.1, respectively. When one measures these quantities, the observed peak profile, i.e. $I_{obs}(Q)$ versus Q, is always convoluted by an instrumental resolution R as $I_{obs}(Q) = \int I(Q-x)R(x)dx$. If I and R are both Gaussian or Lorentzian, one can obtain the following simple relation for the peak width (usually defined as the full-width at half-maximum value):

$$W_{obs}^2 = W^2 + W_{res}^2 \qquad \text{for Gaussian,} \qquad (2.7)$$

$$W_{obs} = W + W_{res} \qquad \text{for Lorentzian.} \qquad (2.8)$$

From these equations, one can estimate the intrinsic width W after the resolution width W_{res} is measured by using a standard specimen such as an epitaxially-grown perfect superlattice for the small-Q (or called small-angle) region and a perfect Si or Ge crystal for the medium-Q (or called medium-angle) region.

(iii) <u>Criterion for kinematical or dynamical diffraction</u> A Bragg-diffracted x-ray beam is diffracted again when it passes through another part of a crystal which is aligned at the right angle for Bragg reflection. This condition is

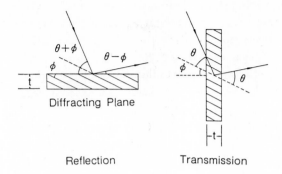

$\theta + \phi$ ϕ $\theta - \phi$

t

Diffracting Plane

θ ϕ θ

$\vdash t \dashv$

Reflection   Transmission

Fig. 2.2.4 Diffraction geometry of x-rays with respect to the diffracting plane and surface of a crystal.

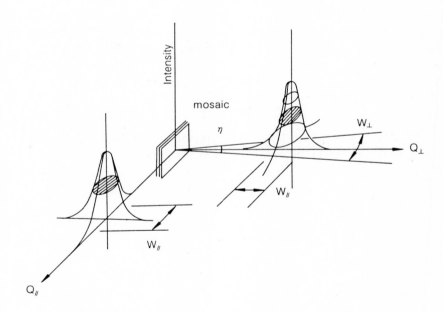

Fig. 2.2.5 Longitudinal ($W_{//}$) and transverse ($W_\perp$) widths measured by x-ray diffraction give information on the coherence length and mosaic ($\eta = W_\perp/Q$) of a superlattice, respectively. $W_{//}$ along $Q_\perp$ is related with the interlayer coherence while the lateral coherence influences $W_{//}$ along $Q_{//}$.

always ensured in a perfect crystal, but it becomes less probable when there are crystal imperfections which deteriorate the periodicity of a set of atomic planes. For a perfect crystal, therefore, multiple scattering processes give rise to the interference so-called "dynamical diffraction"(5). On the other hand, if the atomic planes are not sufficiently ordered to give well-defined diffracted beams, the multiple scattering effect is less important and the single scattering process becomes dominant. An extreme case of this trend is called "kinematical diffraction".

According to the dynamical diffraction theory, the extinction distance t_B is a measure of the depth into a crystal necessary for the multiply-diffracted beams to make a full interference [see Eq.(2.36)]. If the x-ray beams lose their phase relation before they reach t_B due to the following reasons, dynamical diffraction is not fully ensured: (a) X-ray beams are absorbed due to the photo-ionization effect at the depth t_a shorter than t_B. (b) As x-ray beams penetrate into a strained crystal with a gradual change in its lattice-spacing, the Bragg condition is no longer satisfied at the depth t_π shorter than t_B. (c) The thickness of the superlattice t_s is smaller than t_B. However, it should be noted that there is no clear boundary to judge which diffraction process takes place in a given crystal. Maybe the best way to judge is to compare the observed rocking curve of the Bragg reflection with the calculated one based on the dynamical theory as presented in Section 2.4.

2.3   KINEMATICAL DIFFRACTION

The kinematical diffraction theory is the most widely used to analyze the observed intensity data of superlattices. By becoming familiar with this treatment, one can tell the structural properties of a given superlattice from a glance at its diffraction patterns. In this section, we deal with an ideal superlattice first and later introduce imperfections.

2.3.1   Ideal superlattices

Let us first employ a one-dimensional model to describe the layered structure of a superlattice stacked along the z direction. In a perfectly periodic superlattice with its superlattice period Λ and its length (thickness) $N\Lambda$ from $z = 0$ through $z = N\Lambda$ (N = total number of unit layers) as shown in Fig.2.3.1, the intensity is calculated as

$$I(Q) = L_N(Q)|F(Q)|^2 ,$$

(2.9)

where

$$L_N(Q) = [\sin(N\Lambda Q/2)/\sin(\Lambda Q/2)]^2 , \qquad\qquad (2.10)$$

$$F(Q) = \int_0^\Lambda \rho(z) \exp(iQz)dz = \sum_\nu f_\nu(Q) T_\nu(Q) \exp(iQz_\nu) . \qquad (2.11)$$

Here $\rho(z)$ is the electron charge density, $f_\nu(Q)$ is the atomic scattering factor times atomic density in the ν-th plane and Q is the magnitude of the scattering vector defined by Eq.(2.3). The last expression in Eq.(2.11) takes into account the thermal vibration effect as the Debye–Waller factor $T_\nu(Q) = \exp[-B_\nu(Q/4\pi)^2] = \exp[-B_\nu(\sin\theta/\lambda)^2]$ where B_ν is called the temperature factor. The Laue function $L_N(Q)$ has its maximum N^2 at equally-distant positions given by integral multiples of Λ^* and its width $\Gamma_N = 2\pi/N\Lambda$ as shown in Fig.2.3.8. The function

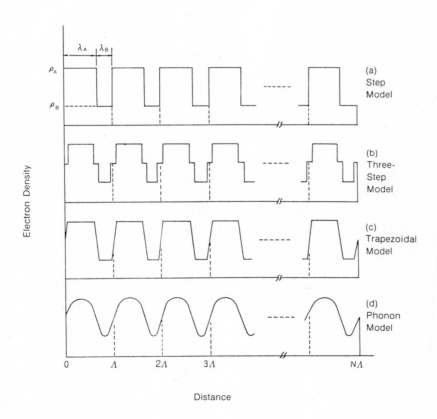

Fig. 2.3.1   One-dimensional superlattice models with a perfectly periodic structure (period Λ). A charge-density contrast is shown schematically for N bilayers with four kinds of interfacial structures (a)-(d).

F(Q) is a layer structure factor reflecting the structure within a unit layer. For example, the structure factors for four models with different chemical ordering schemes depicted in Fig.2.3.1 are different but their Laue functions are identical. That is, the equally-distant Bragg peaks have their intensities modified by the intralayer structure(6).

(i) <u>Intralayer order</u> In order to carry out further calculations of F(Q) in Eq.(2.11), we introduce a step model which assumes a complete alternation of chemical compositions at interfaces between adjacent layers A and B with thicknesses λ_A and λ_B, respectively. In this case, Eq.(2.11) is rewritten as

$$|F(Q)|^2 = |\int_0^{\lambda_A} \rho(z) \exp(iQz)dz + \int_{\lambda_A}^{\lambda_A+\lambda_B} \rho(z) \exp(iQz)dz|^2$$

$$= f_A^2 L_A(Q) + f_B^2 L_B(Q) + 2f_A f_B [L_A(Q)L_B(Q)]^{1/2}\cos(\Lambda Q/2), \quad (2.12)$$

where $\Lambda = \lambda_A + \lambda_B$. The function $L_\nu(Q)$ can be calculated as

$$L_\nu(Q) = \begin{cases} [\sin(\lambda_\nu Q/2)/(\lambda_\nu Q/2)]^2 & \text{for an amorphous layer,} \quad (2.13) \\ [\sin(n_\nu d_\nu Q/2)/\sin(d_\nu Q/2)]^2 & \text{for a crystalline layer.} \quad (2.14) \end{cases}$$

Here n_ν and d_ν are the number of atomic planes and lattice spacing, respectively. In the derivation of Eq.(2.14), the interface thickness is assumed to be $(d_A + d_B)/2$, thereby $\Lambda = n_A d_A + n_B d_B$. Eq.(2.13) is a function monotonically decreasing with increasing Q while Eq.(2.14) is the Laue function resulting from periodically aligned atomic planes within the crystalline layer. The latter function has its peaks at positions expressed by integral multiples of $d_\nu^* = 2\pi/d_\nu$ and width $\Gamma_\nu = 2\pi/n_\nu d_\nu$. It corresponds to the structure factor for the bulk structure, which gives rise to the Bragg reflection in the medium-Q region. In contrast the Laue function given by Eq.(2.10) originating from the superlattice period shows its low-order reflections in the small-Q region. When both layers are crystalline, the structure factor of Eq.(2.12) shows its peak at a position shifted from the bulk reflection because of the interference term in the same equation. Figure 2.3.2 displays the intensity calculated by Eq.(2.9) for each combination of amorphous and crystalline layers. The dashed curve represents $|F(Q)|^2$. One can see how diffraction patterns change depending upon the intralayer structure. There is no significant difference in the patterns in the small-Q region, which result only from a contrast of the electron charge density between two kinds of layers. Thus the diffraction pattern in the medium-Q region gives direct information on the intralayer structure, i.e. amorphous or crystalline, and indicates which atomic planes are stacked in the growth direction.

In Fig.2.3.2 one can see that all peaks appear at positions expressed by integral multiples of Λ^* because $\Gamma_N \ll \Gamma_A$, Γ_B due to the fact that $\Lambda \gg d_A$, d_B. Here it should be noted that an observed peak shifted from the bulk position does not necessarily imply the presence of strain in the crystalline layer.

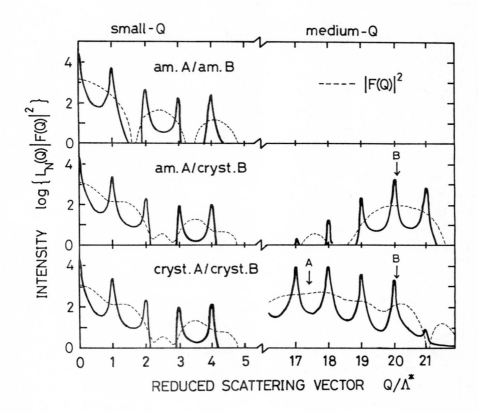

Fig. 2.3.2 *Step-model calculations demonstrating how differently diffraction patterns appear depending upon the intralayer structure of layers A and B. The crystalline and amorphous states are denoted by the abbreviations of "cryst" and "am". Solid and dashed curves drawn as a function of a reduced scattering vector represent intensity and structure factor squared in a logarithmic scale, respectively. In these figures, only the envelope of the oscillatory intensity curve governed by the Laue function is drawn neglecting the details of subsidiary peaks. The layer thickness is fixed as λ_A(amorphous) = $n_A d_A$(crystalline) = $7 \times 2.68\text{Å}$ and λ_B(amorphous) = $n_B d_B$(crystalline) = $12 \times 2.32\text{Å}$. Other parameters are also fixed as $f_A = 26$, $f_B = 12$ and $N = 12$. Arrows with symbols A and B show the positions of the bulk reflections given by $d_A^* = 2\pi/d_A$ and $d_B^* = 2\pi/d_B$, respectively.*

This is clearly shown in Fig.2.3.3 where it is demonstrated how the diffraction pattern from the intralayer bulk structure is affected by the relative magnitude of the lattice-spacings d_A and d_B. In the case of $d_A = d_B$ as shown in Fig.2.3.3(d), one can obtain $\Lambda = n_A d_A + n_B d_B = (n_A + n_B)d$, i.e. $d^* = n\Lambda^*$ (n; integer). That is, the peak position of $|F(Q)|^2$ is very close to that of $L_N(Q)$. This fact results in a strong bulk diffraction peak surrounded by its accompanying satellites. The diffraction patterns experimentally observed on [Fe(15Å)/Mg(30Å)]$_{60}$(6) and [Fe(15Å)/V(30Å)]$_{60}$(7) are very similar to those of Figs.2.3.3(a) and (c), respectively. The former has a textured intralayer structure of Fe(110)$_{bcc}$/Mg(00.1)$_{hcp}$ [$d_{Fe} = 2.027$Å, $d_{Mg} = 2.605$Å] while the latter Fe(110)$_{bcc}$/V(110)$_{bcc}$ [$d_V = 2.149$Å]. A pattern similar to Fig.2.3.3(c) or (d) is also obtained for an epitaxially-grown superlattice in which the lattice-spacing is coherently relaxed to accomodate strain. It is shown later how to distinguish between these two cases.

When the layer thickness Λ becomes larger, more diffraction peaks appear in both small- and medium-Q regions and a strong bulk reflection approaches closer to the bulk position as shown by model calculations in Fig.2.3.4. Such a systematic change in the diffraction pattern in the medium-Q region has been actually observed in several superlattices such as Au/Pd(8) and Mo/Ni(9).

To summarize this subsection: (a) The superlattice period can be measured from the peak position in the small-Q region. However, since a low-angle peak position is affected by refraction as discussed in 2.4.2, it is recommended to use the position of the higher-order reflections or distances between them. (b) In the case of a crystalline layer, Λ can also be obtained from the distance between the satellite peaks about the bulk reflection observed in the medium-Q region. (c) The stacking direction of atomic planes in the crystalline layer can be determined from observing the bulk reflections.

(ii) <u>Interlayer order</u> The width of the Bragg peak is determined by the total number of unit layers N as given by Eq.(2.10). In practice N can be understood as the number of unit layers which diffract x-ray beams coherently. In other words, x-rays diffracted at the first and N-th layers become out of phase because of the irregularity in layer thickness. Thus the peak width provides means for judging how perfectly a periodic structure is established in a superlattice. In order to experimentally estimate N, however, correction for instrumental resolution must be carried out for the observed width as mentioned in 2.2.3 [Eq.(2.7) or (2.8)]. Figure 2.3.5 displays the intensity calculated for several values of N. The width in this figure is normalized as $\Gamma_N/\Lambda^* = 1/N$. Thus the coherence length along the stacking direction is obtained from the observed intrinsic width $W_{//}$ as $\xi_\perp = N\Lambda = 2\pi/W_{//}$.

(iii) <u>Lateral order</u> In order to measure the lateral order, one must examine whether any reciprocal lattice point distant from the principal c^* axis is a

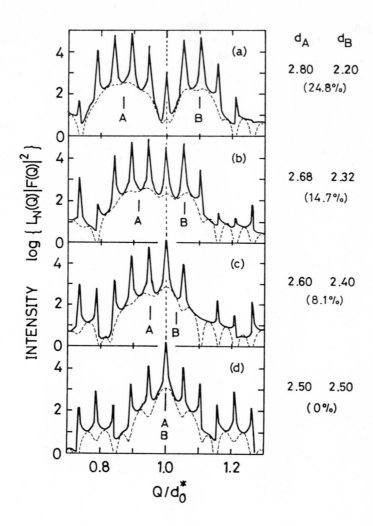

Fig. 2.3.3 Step-model calculations demonstrating the variations in the diffraction patterns in the medium-Q region of a textured superlattice depending upon the relative magnitude of lattice-spacings d_A and d_B. The figure in parenthesis is a measure of the mismatch given by $(d_A - d_B)/d_0$ where the average spacing $d_0 = (n_A d_A + n_B d_B)/(n_A + n_B)$. Solid and dashed curves represent intensity and structure factor squared in a logarithmic scale, respectively. The abscissa is expressed by a reduced unit defined as $Q/d_0^* = Q/(2\pi/d_0)$. Two vertical bars A and B in each figure show the positions of the bulk reflections. Other parameters are fixed as $n_A = 7$, $n_B = 12$, $f_A = 26$, $f_B = 12$ and $N = 12$.

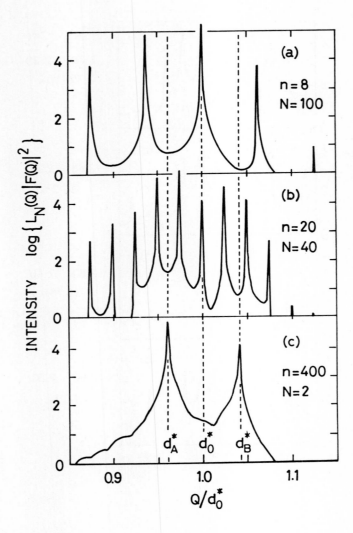

Fig. 2.3.4 Step-model calculations demonstrating the variations in the diffraction patterns in the medium-Q region of a textured superlattice depending upon the layer thickness. As the thickness increases ($n = n_A = n_B$), strong Bragg peaks move from the average position d_0^* towards the bulk positions d_A^* and d_B^*. Other parameters are fixed as $d_A = 2.6\text{Å}$, $d_B = 2.4\text{Å}$, $f_A = 26$, $f_B = 12$ and the thickness $t_s = N\Lambda = 4000\text{Å}$.

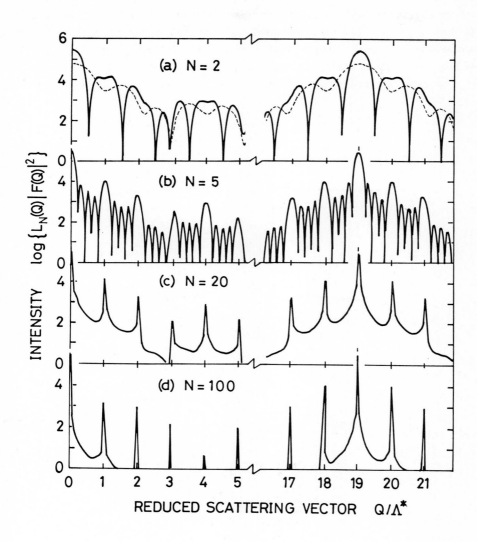

Fig. 2.3.5 Step-model calculations demonstrating how Bragg peaks sharpen as the total number of stacking bilayers N increases. In (c) and (d), only the envelope of the intensity curve is drawn neglecting the details of the subsidiary peaks of the Laue function. Dashed curves in (a) show the structure factor squared. Other parameters are fixed as $d_A = d_B = 2.5$Å, $n_A = 7$, $n_B = 12$ ($\Lambda = 47.5$Å), $f_A = 26$ and $f_B = 12$.

52

spot or ring as shown in Fig.2.2.1. One useful method is to take photographs
similar to the Laue or oscillation method by illuminating a superlattice with
white or monochromatic x-rays, respectively. In the latter case, the specimen
must be oscillated during exposure. When the photo shows spots, the
superlattice is a good single crystal as a whole indicating epitaxial growth.
When the photo shows continuous arcs or rings, on the other hand, it indicates
that the superlattice is comprised of many randomly-oriented grains in the

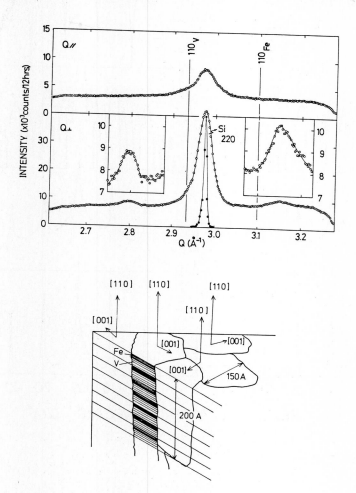

Fig. 2.3.6 The broad longitudinal widths of the 110 reflection in
the medium-Q region observed along $Q_{//}$ and $Q_\perp$ on $[Fe(15\text{Å})/V(30\text{Å})]_{60}$.
The instrumental resolution was obtained from the width of the Si
220 standard reflection as shown. The coherence length estimated
from Eq.(2.7) suggests a domain structure shown schematically at
the bottom. [From Endoh et al.(7)]

lateral direction. By using this technique, Falco et al. and Greene et al.(10) studied the lateral ordering in Ta/Mo and Nb/Er superlattices, respectively. In the latter case, it was observed to be significantly influenced by the superlattice period.

In order to examine the lateral order more quantitatively, the diffractometer method is used to measure the longitudinal width of any reciprocal lattice point or arc in the a^*b^* plane perpendicular to the growth direction. The resolution-corrected width W gives a lateral coherence length of $\xi_{//} = 2\pi/W_{//}$. In this method, the superlattice as well as the substrate must be capable of transmitting x-rays. It should be noted that the other way to obtain $\xi_{//}$, particularly for a superlattice incapable of transmitting x-rays, is to measure the peak width along $Q_{//}$ of any reciprocal lattice point which can be reached in the (asymmetric) reflection geometry as shown in Fig.2.2.4. The top and bottom figures of Fig.2.3.6 show the measurements of both $\xi_{//}$ and $\xi_\perp$ for $[Fe(15Å)/V(30Å)]_{60}$, respectively(7). The dots represent the instrumental resolution estimated by the 220 reflection of a standard Si powder specimen. By using Eq.(2.7), the coherence length obtained was $\xi_{//} = 150$ Å and $\xi_\perp = 200$ Å. Based on these values and the fact that randomly-oriented grains are present in the lateral direction, the structure of this superlattice is shown schematically in the same figure. A recent observation of the same superlattice by electron-microscope confirmed such a grain size. Figure 1.3.3 in Chapter 1 shows a sectional micrograph of $[Fe(4Å)/Mg(16Å)]_{200}$, which also has a similar randomly-oriented grain distribution(11). One can see that both interlayer and lateral coherences are limited by dislocations.

(iv) <u>Interface and strain</u> An interface is a matter of primary interest in superlattices because of its potential capability to create novel solid state properties which do not exist in nature. From a structural point of view, the compositional gradient and atomic positions in the interfacial region are subject to study. The four interface models of Fig.2.3.1 are typical examples of compositional distribution. However, it is difficult to identify the interface structure from a glance at the observed diffraction patterns. In general the steeper compositional gradient produces the stronger higher-order reflections. This results from the fact that the number of observable reflections is determined by the number of Fourier components necessary for describing a compositional wave, as given by the inverse Fourier transform of Eq.(2.11). The step model [Fig.2.3.1(a)] produces an infinite number of reflections while an ideal sinusoidally-modulated phonon model (a special case of Fig.2.3.1(d)) gives only the first-order reflection. It is reported that there is a tendency of the compositional gradient to be steeper in a superlattice prepared at a lower temperature. Usually the shape of the compositional wave is obtained by model-fitting to the observed intensities by adjusting some parameters such as interface thickness and slope of the compositional gradient. If one can measure accurately the integrated intensity of reflections along the $Q_\perp$ direction, the Fourier transform of the layer structure factor [Eq.(2.11)] can provide the charge density distribution, i.e. compositional wave. However, as always encountered in a process of a conventional structure analysis, the phase problem still remains unsolved. The four interface models of Fig.2.3.1 are realized approximately in the following cases:

MODEL	ELEMENTAL COMBINATION	SUPERLATTICES
(a) step model	insoluble	Fe/Mg, W/C, W/Si
(b) three-step model	intermetallic compound	Mn/Sb, Co/Sb
(c) trapezoidal model	soluble	Nb/Al
(d) phonon model	soluble	Cu/Ni, Nb/Ta, Gd/Y

The three-step model is used to describe the interface of an intermetallic compound. In the case of Mn/Sb, Mn reacts with Sb to form MnSb which is known to be stable in the bulk alloy. Figure 1.4.1 in Chapter 1 shows the observed diffraction pattern in the medium-Q region of $[Mn(1Å)/Sb(60Å)]_{40}$, which is well explained by this model with a monolayer-thick compound interface of MnSb(12). The trapezoidal model was used to analyze the data of the Nb/Al superlattice by McWhan et al.(13). The phonon model is suitable for an epitaxially-grown superlattice in which any mismatch between layers due to different atomic sizes is accomodated by elastic strains. In metallic superlattices, successful epitaxial growth has been reported for a few cases such as Cu/Ni(14), Nb/Ta(15) and Gd/Y(16).

In order to see the strain effect on the diffraction pattern, let us employ the one-dimensional phonon model calculated first by Esaki et al.(17) and later extended by Segmüller and Blakeselee(18). Although McWhan cited their results in his book(1), we also describe the essence of their calculations for the reader's convenience. The lattice (interplanar) spacing between the (j-1)-th and j-th planes is modulated by strains and may be described by the Fourier sum as

$$d_j = d_0[1 + \sum_{p=1}^{\infty} \Delta_p \cos(p\Lambda^* z_j)] , \qquad (2.15)$$

where d_0 is the average lattice-spacing and Δ_p is the amplitude of the p-th harmonic of the strain wave. The scattering power is also modulated by atomic interdiffusion and it may be given by

$$f(z_j) = f_0[1 + \sum_{p=1}^{\infty} \phi_p \cos(p\Lambda^* z_j)] , \tag{2.16}$$

for the j-th plane at the position z_j. Here f_0 is the average scattering power. Since the position of the j-th plane in the m-th layer is given by

$$z_{mj} = (m-1)\Lambda + z_j = (m-1)\Lambda + \sum_{k=1}^{j-1} d_k , \tag{2.17}$$

the intensity is calculated as $I(Q) = L_N(Q)|F(Q)|^2$ where L_N is the same Laue function as in Eq.(2.10) and

$$F(Q) = \sum_{j=1}^{n} f(z_j) \exp(iQ \sum_{k=1}^{j-1} d_k) . \tag{2.18}$$

Here n is the total number of planes present in the unit layer, i.e. $\Lambda = nd_0$. The exponent in Eq.(2.18) can be expanded in a series of Bessel functions J_p, but the expression becomes so complex that one cannot easily use it to fit the experimental data. By truncating the expansion of both the Bessel function and Fourier sum to terms up to the second-order, one can obtain such convenient expressions as the following:
For the main Bragg peak $(Q = \ell d_0^*$ with integer ℓ),

$$F(\ell) = f_0[J_{01}J_{02} - \phi_1 J_{11}J_{12} + \phi_2 J_{21}(J_{02} + J_{22})] . \tag{2.19}$$

For the first-order satellites $[Q = (\ell \mp 1/n)d_0^*]$,

$$F(\ell \mp 1/n) = f_0\{ \pm J_{11}J_{02} - J_{11}J_{12}$$

$$+ \phi_1[J_{01}(J_{02} \mp J_{12}) + J_{21}(J_{02} + J_{22})]/2$$

$$\mp \phi_2 J_{11}(J_{02} + J_{22})/2\} . \tag{2.20}$$

For the second-order satellites $[Q = (\ell \mp 2/n)d_0^*]$,

$$F(\ell \mp 2/n) = f_0\{ \pm J_{01}J_{12} + J_{21}(J_{02} + J_{22})$$

$$\pm \phi_1 J_{11}(J_{02} - J_{22})/2$$

$$+ \phi_2[J_{01}(J_{02} + J_{22}) \pm J_{21}J_{12}]/2\} , \tag{2.21}$$

56

where $J_{p1} = J_p(\Delta_1 n\ell)$ and $J_{p2} = J_p(\Delta_2 n\ell/2)$. From a graph of the Bessel functions, one can easily see that the strain causes an asymmetry in satellite intensities normalized by f_0^2 with respect to the main peak. The step model introduced in 2.3.1 can be described by an infinite number of Fourier components of both compositional and strain waves. This corresponds to a special case of the phonon model which produces an infinite number of satellites. In an actual three-dimensional lattice, the coherency strain built along the growth direction accompanies a lateral strain in the film plane

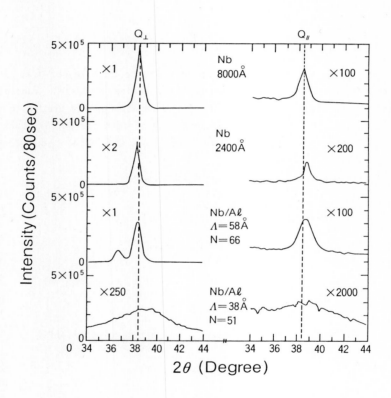

Fig. 2.3.7 Diffraction patterns observed along $Q_\perp$ and $Q_{//}$ of equi-
layered Nb/Al superlattices and pure Nb metal deposited on the (100)
surface of an oxidized silicon. The vertical dashed lines indicate
the position of the 110 reflection of the bulk Nb which is only in
agreement with that observed on a thick (8000Å) film of pure Nb. A
thinner (2400Å) film of Nb shows an orthorhombic distortion with
the lattice constant along $Q_\perp$ expanded and along $Q_{//}$ contracted.
Such a distortion is also observed on the superlattice with $\Lambda = 58Å$
which demonstrates the coherency strain. A loss of texture
$(110)_{Nb}/(111)_{Al}$ and a decrease in grain size are also seen for $\Lambda = 38Å$. [From McWhan et al.(13)]

according to the Poisson ratio. For example, in the superlattice [Nb(29Å)/Al(29Å)]$_{66}$(13) with its small lattice mismatch [d$_{Nb(110)}$ − d$_{Al(111)}$]/d$_0$ = 0.17%, its main (bulk) Bragg reflection appears at a diffraction angle slightly smaller than that of pure Nb along Q$_\perp$ as shown in Fig.2.3.7. In the film plane Q$_{//}$, on the other hand, the average Bragg peak position shifts towards the higher angle. Along the stacking direction, therefore, the lattice spacing of the Nb(110) plane is expanded with respect to the pure element and conversely, the Al(111) lattice spacing is contracted. McWhan et al. also confirmed this fact by observing the asymmetry in the satellite intensities as treated above. Recently Miceli et al.(19) observed a strain modulation in a Nb/Ta superlattice induced by absorbed hydrogens. Their analysis of the asymmetry in the intensities of a pair of satellites revealed that the Nb absorbs more hydrogen than the Ta as observed in the bulk while hydrogen solubility increases remarkably in the superlattice.

In summary of this subsection, the coherency strain can be examined experimentally in two ways: (a) The average main Bragg reflection appears at a position shifted from that for the pure element conversely in the Q$_\perp$ and Q$_{//}$ directions. (b) A pair of satellites of the same order about the main peak is asymmetric in their intensities.

2.3.2 Imperfect superlattices

In the process of preparing superlattices, fluctuation of the layer thickness may be introduced by inadequate instrumental controllability or any other reasons. Here we follow the calculation by Fujii et al.(6), who treated this problem as a concept of "paracrystalline disorder". When such a fluctuation in the superlattice period Λ is expressed by the Gaussian distribution exp[−(Λ−Λ$_0$)^2/ σ^2] (non−cumulative disorder), the Laue function Eq.(2.10) is modified as

$$\tilde{L}_N(Q) = \frac{1 + \exp(-N\sigma^2 Q^2/2) - 2\exp(-N\sigma^2 Q^2/4)\cos(N\Lambda_0 Q)}{1 + \exp(-\sigma^2 Q^2/2) - 2\exp(-\sigma^2 Q^2/4)\cos(\Lambda_0 Q)} \quad , \qquad (2.22)$$

where Λ$_0$ is the average of the superlattice period. Here a random distribution of different superlattice periods is assumed to derive Eq.(2.22). In the limit of σ→0, this equation becomes identical to Eq.(2.10). Figure 2.3.8 displays the behavior of the modified Laue function Eq.(2.22) for N = 12. The degree of fluctuation is given by the full−width at half−maximum relative to Λ$_0$, i.e. ΔΛ/Λ$_0$ = 2√ln2σ/Λ$_0$ = 1.67σ/Λ$_0$. One can see a remarkable reduction of its peak value with increasing Q. On the other hand, the layer structure factor Eq.(2.11) has the fluctuation effect through the position coordinate in its

58

exponent. With the step model, it is calculated in two different ways depending upon the origin of the fluctuation. (a) When the numbers of A and B layers fluctuate and they are expressed by the Gaussian distribution $\exp[-(n_\nu - n_{\nu 0})^2/\sigma_\nu^2]$ ($\nu = A, B$), we obtain

$$|F(Q)|^2 = f_A^2 L_A(Q) \exp(-d_A^2 \sigma_A^2 Q^2/8) + f_B^2 L_B(Q) \exp(-d_B^2 \sigma_B^2 Q^2/8)$$

$$+ 2f_A f_B [L_A(Q) L_B(Q)]^{1/2} \times \exp[-(d_A^2 \sigma_A^2 + d_B^2 \sigma_B^2)Q^2/16]\cos(\Lambda_0 Q/2) .$$

(2.23)

Here the Laue functions L_A and L_B are given by Eq.(2.14) by putting $n_A=n_{A0}$ and

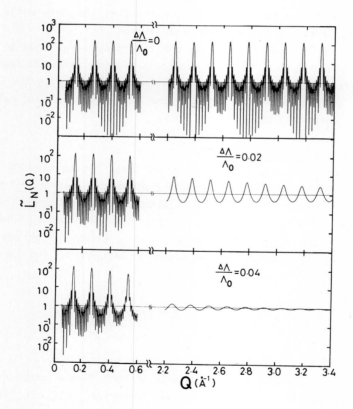

Fig. 2.3.8 *Modified Laue function with a fluctuation in superlattice period [Eq.(2.22)] displays an intensity reduction as the scattering vector increases. The following parameters which correspond to A = (110)Fe and B = (00.2)Mg are used in these calculations:* d_A = 2.027Å, d_B = 2.605Å, n_A = 8, n_B = 12 (Λ_0 = 47.48Å, Λ_0^* = 0.1323Å^{-1}) *and N = 12. The top figure corresponds to the ideal Laue function given by Eq.(2.10). [From Fujii et al.(6)]*

$n_B = n_{B0}$.   (b) When the interface thickness fluctuates as $\exp[-(d_t - d_{t0})^2/\sigma_t^2]$ around $d_{t0} = (d_A + d_B)/2$, we obtain

$$|F(Q)|^2 = \text{Eq.}(2.12) \times \exp(-\sigma_t^2 Q^2/8) .$$ (2.24)

The parameter σ in Eq.(2.22) is expressed as $2n_\nu\sigma_\nu$ in case (a) and $2\sigma_t$ in (b). In both cases, the fluctuations play the role of an effective Debye-Waller factor in $F(Q)$, which reduces the peak intensity but doesn't affect the peak width. However, the overall intensity profile governed by $L_N(Q)|F(Q)|^2$ shows a remarkable peak broadening as Q increases. This results from the fact that with increasing fluctuation the shape of $|F(Q)|^2$ manifests itself more apparently in the intensity profile because $L_N(Q)$ tends to 1.   Figure 2.3.9 demonstrates such a behavior.   The peak width in the small-Q region is determined by N, which is the number of unit layers coherently stacked in a sense of x-ray diffraction. On the other hand, the fluctuation causes peak-broadening and intensity-

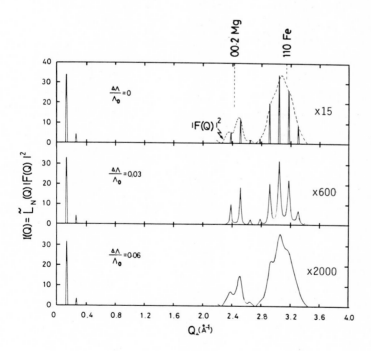

Fig. 2.3.9 *Overall diffraction patterns calculated on an Fe/Mg superlattice with the $(110)_{Fe}/(00.1)_{Mg}$ texture for various magnitudes   of fluctuation. The dashed curve in the top figure represents   the structure factor squared in the medium-Q region. The two vertical bars show the positions of the 00.2_{Mg} and 110_{Fe} bulk reflections. The parameters used are the same as those for Fig.2.3.8. [From Fujii et al.(6)]*

reduction, which appear significantly in the medium–Q region.   The bottom figure of Fig.2.3.10 shows a similar calculation for $\Delta\Lambda/\Lambda_0 = 0.054$ and N = 12, which is the best fit to the diffraction pattern   observed for [Fe(15Å)/Mg(30Å)]$_{60}$ shown at the top of the same figure(6).   For a direct comparison with the observation, the calculation took into account the convolution of an instrumental resolution, the width of which was experimentally determined as drawn by horizontal bars in the figure (a).   The observed peak-broadening and intensity–reduction can be well explained by this model and the 5.4 % fluctuation in Λ corresponding to about one layer thickness is likely to exist in this superlattice.   A similar fluctuation was found in many other superlattices such as Pd/Au(8), Nb/Al(13) and Pb/Ag(20).

The superlattice [Fe(15Å)/Mg(30Å)]$_{60}$ was also used to study the thermal annealing effect as briefly introduced in Chapter 1, where observed diffraction patterns as a function of annealing time at 155°C are shown in Fig.1.5.2.   As annealing time increased up to 52 hours, the small–Q Bragg reflections became stronger while all peaks in the medium–Q region became sharper.   The model-fitting to these data revealed a significant improvement of the interlayer order by thermal annealing(21).   Beyond 52 hours, however, all peaks started

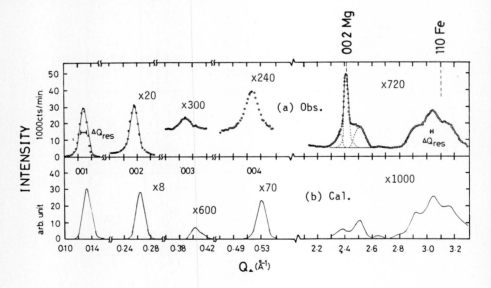

Fig. 2.3.10   (a) Overall diffraction pattern observed along $Q_\perp$ of [Fe(15Å)/Mg(30Å)]$_{60}$. (b) The best fitted pattern by the resolution-convoluted model calculation with N = 12 and $\Delta\Lambda/\Lambda_0 = 0.054$. Notice the change in Q-scale between the small- and medium-Q regions. Q_{res} shows the instrumental resolution. The strong 00.2 reflection of Mg comes from a thick (200Å) cover layer deposited on this superlattice. [From Fujii et al.(6)]

broadening and those measured at 194 hours showed no further appreciable change. It was also observed that the annealing temperature of 250°C resulted in the complete destruction of the superlattice structure within several hours. This kind of thermal annealing experiment to study the thermal stability of superlattices under heat-loading is very important for superlattices which are designed as optical devices. Dinklage(22) first carried out such a measurement on several superlattices made for soft x-ray monochromators.

2.4   DYNAMICAL DIFFRACTION

The application of the dynamical diffraction theory to superlattices is well explained by Underwood and Barbee(23). This section is based on their formulation. However, some quantities are expressed differently to conform to the notations   used in the preceding sections.

2.4.1 Total reflection

Most materials have a refractive index slightly less than unity for x-rays. Consequently x-rays incident on the material are totally reflected when the glancing angle is smaller than the critical angle given below. The complex index of refraction is described by

$$n^* = 1 - \delta - i\beta ,$$ (2.25)

where δ and β for elemental materials can be expressed as

$$\delta = N_0 r_e \lambda^2 (Z + \Delta f')/2\pi ,$$ (2.26)

$$\beta = N_0 r_e \lambda^2 \Delta f''/2\pi = \mu\lambda/4\pi .$$ (2.27)

Here $r_e = e^2/mc^2 = 2.818 \times 10^{-13}$ cm, Z the atomic number, N_0 the number density of atoms and μ the linear absorption coefficient. The critical angle for the total reflection is calculated as

$$\theta_c = \sqrt{2\delta} = 2.99 \times 10^{-23}\sqrt{n_e}\lambda ,$$ (2.28)

where θ_c, n_e and λ are given in units of radians, electrons/cm^3 and Å, respectively. Since the total reflection occurs at a very small angle, for example, $\theta_c = 10$mrad $= 0.57°$ for Au at $\lambda = 1.54$Å, it can be measured only with incident x-ray beams collimated well enough to separate the direct and reflected beams.

A special grazing—incidence diffraction technique developed by Marra et al.(24) exploits the total reflection for eliminating any diffraction from the substrate, on which the target layer to be investigated is deposited or adsorbed. A weak signal concerning the in-plane structure of the deposited layer can be detected. In this case, the critical angle of the substrate must be larger than that of the deposited layer. This technique is very promising for studying the structure of a thin superlattice.

2.4.2 Bragg reflection

When the glancing angle becomes larger than the critical angle, x-rays penetrate into the material. Then they are Bragg—diffracted by a certain set of diffracting planes, which in superlattices corresponds to a set of parallel interfaces. In a perfectly—periodic superlattice, a Bragg—diffracted beam is diffracted again when it passes through another part of the superlattice. Such a multiple diffraction process can be treated either by an analytical method or by a computational one. The former method becomes very complex when one considers the absorption while the latter is readily available in any case. In this subsection, we first describe the analytical formula for a zero-absorption case to obtain the physical insight into the dynamical diffraction in superlattices. Then the computational method is presented for a general case. Here we consider only diffraction in the small—Q region.

(i) Absorption—free case Among various analytical treatments of dynamical diffraction, let us employ the Ewald formulation used by Underwood and Barbee(23), which is adequate for illustrative purpose. When x-rays with the incident intensity I_0 are diffracted symmetrically as shown in Fig.2.4.1 (Bragg case), the diffracted intensity as a function of the glancing angle θ is given by

$$R = I(\theta)/I_0 = [y^2 + (y^2 - 1)\cot^2(A\sqrt{y^2 - 1})]^{-1} \qquad (2.29)$$

where

$$A = 4\pi r_e N_0 PN|F_{\mathbf{H}}'|/H , \qquad (2.30)$$

$$y = \frac{k^2}{2Pr_e N_0 \Lambda^* |F_{\mathbf{H}}'|} [(\theta - \theta_{\mathbf{H}})\sin2\theta_{\mathbf{H}} - 2\delta_0] , \qquad (2.31)$$

$$\delta_0 = \frac{2\pi r_e}{k^2} \sum n_j(f_j^0 + \Delta f_j') . \qquad (2.32)$$

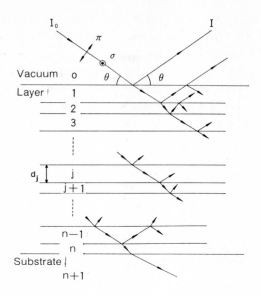

Fig. 2.4.1 *Schematic view of x-rays being scattered and transmitted at each interface of a superlattice composed of n layers. The symbols* σ *and* π *denote the polarization direction of incident x-rays as shown.*

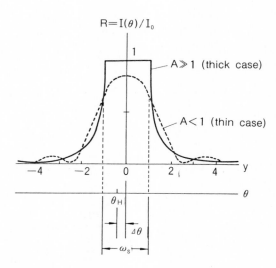

Fig. 2.4.2 *A schematic drawing of the rocking curve (reflectivity versus glancing angle) of Bragg reflection in the Bragg case (reflection geometry).*

F_H' is the real part of the layer structure factor for the reciprocal lattice point H given by Eq.(2.11), n_j the number density of the j-type atoms averaged over the structure, P the polarization factor given as 1 for the σ- and $|\cos2\theta_H|$ for the π-components, and θ_H the kinematical Bragg angle satisfying $2k\sin\theta_H = H$. Figure 2.4.2 illustrates schematically the behavior of Eq.(2.29) called a "rocking curve". It has a high reflectivity in the range of $-1 \leq y \leq 1$. This range converted to the angle is called the "intrinsic width" or "Darwin width" which is calculated as

$$\omega_s = \theta(y=1) - \theta(y=-1) = \frac{4Pr_eN_0\Lambda^*|F_H'|}{k^2\sin2\theta_H} \quad . \tag{2.33}$$

This quantity is usually of the order of several seconds of arc. The peak intensity, i.e. the maximum reflectivity, is obtained at $y = 0$ as

$$R_{max} = [I(\theta)/I_0]_{y=0} = \tanh^2A \quad . \tag{2.34}$$

This quantity is relevant to evaluate the reflectivity of a superlattice applied to a monochromator. For $A \gg 1$ (thick case), $R_{max} = 1$ while $R_{max} = A^2$ for $A < 1$ (thin case). The former refers to a superlattice thick enough for x-rays to make a full interference while the latter is for a thin superlattice in which the multiple diffraction is not accomplished. It is also shown that the integrated intensity is proportional to A, i.e. $|F_H|$, for the thick case and to A^2, i.e. $|F_H|^2$, for the thin case. The latter case corresponds to the kinematical diffraction described in Section 2.3 while the former is

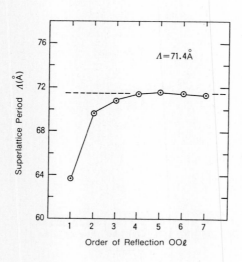

Fig. 2.4.3 Superlattice period of $[Fe(15\text{Å})/Mn(50\text{Å})]_{40}$ directly calculated from the observed position of the 00ℓ reflections in the small-Q region. This calculation is based on the kinematical Bragg's law and neglects the refraction effect.
[From Nakayama et al.(4)]

characteristic of the dynamical diffraction. From Eq.(2.31), one can also notice that the central position (y = 0) of the rocking curve θ_p is not equal to the kinematical Bragg position θ_H because of the refraction effect. Since their difference is given by

$$\Delta\theta = \theta_p - \theta_H = 2\delta_0/\sin 2\theta_H , \qquad (2.35)$$

a lower order reflection is more influenced by this effect. Thus one should be careful to deduce the superlattice period Λ from the observed peak position $\theta_{obs} = \theta_p$ for a lower order reflection. Figure 2.4.3. shows Λ directly determined by $\Lambda = \ell\lambda/2\sin\theta_{obs}$ for the ℓ-th order reflection of the Fe/Mn superlattice(4).

It is difficult to judge which treatment, kinematical or dynamical diffraction, is adequate for a given superlattice. A measure for the thickness to assure a full interference of x-rays is the "extinction distance" defined as

$$t_B = \frac{V}{8\pi r_e} \frac{H}{|F_H'|} . \qquad (2.36)$$

When the superlattice thickness $t_s = N\Lambda$ is larger than t_B, the diffraction process must be treated by dynamical diffraction. As seen in Eq.(2.36), the extinction distance becomes shorter for a stronger reflection so that the diffraction process becomes more dynamical. This is known as the "primary extinction effect" in kinematical diffraction. In the case of [Fe(15Å)/Mg(30Å)]$_{60}$ studied by Fujii et al.(6), for example, t_B for the first order reflection is calculated to be about 1200 Å for $\lambda = 1.54$ Å while nominally $t_s = 2700$ Å. However, its rocking curve was measured to be wide enough to justify the use of the kinematical treatment, which resulted in the coherence length $\xi_\perp = 540$ Å [= t_π as introduced in 2.2.3(iii)] limited by the irregularity in its superlattice period.

(ii) General case A general case of superlattices with absorption and even with different layer thicknesses can be readily treated by a computational method based on the Fresnel equations. Since details of this method are described in the paper by Underwood and Barbee(23), several formulas necessary for the calculations are given here. The reflection coefficients of x-rays at the interface between the j-th and (j+1)-th layers (see Fig.2.4.1) are given as a recursion relation

$$r_{j,j+1} = a_j^4 \frac{r_{j+1,j+2} + f_{j,j+1}}{r_{j+1,j+2} f_{j,j+1} + 1} , \qquad (2.37)$$

where the Fresnel coefficient

$$
f_{j,j+1} = \begin{cases} \dfrac{g_j - g_{j+1}}{g_j + g_{j+1}} & \text{for } \sigma \text{ polarization,} \\[3mm] \dfrac{g_j/n_j^{*2} - g_{j+1}/n_{j+1}^{*2}}{g_j/n_j^{*2} + g_{j+1}/n_{j+1}^{*2}} & \text{for } \pi \text{ polarization,} \end{cases} \tag{2.38}
$$

$$
a_j = \exp(-i\,\frac{\pi}{\lambda}\,g_j d_j) , \tag{2.39}
$$

$$
g_j = (n_j^{*2} - \cos^2\theta)^{1/2} \simeq (\theta^2 - 2\delta_j - 2i\beta_j)^{1/2} . \tag{2.40}
$$

The calculation starts from a substrate which is assumed to have an infinite

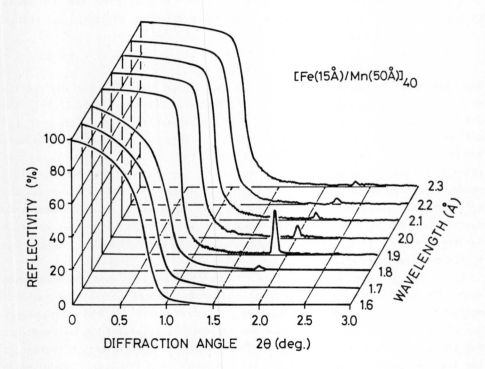

Fig. 2.4.4   Calculated reflectivity curves of the totally- and Bragg- reflected x-rays from [Fe(15Å)/Mn(50Å)]$_{40}$ as a function of wavelength across the K-edge [$\lambda_K(Fe) = 1.743A$ and $\lambda_K(Mn) = 1.896A$]. The remarkable enhancement of the Bragg intensity and a sharp change of the total reflection at the critical angle are seen at $\lambda = 1.900Å \geqslant \lambda_K(Mn) > \lambda_K(Fe)$. This calculation can be directly compared with the observation shown in Fig.2.2.3. [From Nakayama et al.(4)]

thickness and $r_{n,n+1} = 0$. By solving Eq.(2.37) recursively backward to the surface, the relevant reflectivity to be observed is obtained as

$$R = I(\theta)/I_0 = |r_{01}|^2 . \tag{2.41}$$

One may notice that this treatment includes both the total reflection and Bragg reflection. Figure 2.4.4 demonstrates an example of such a calculation for $[Fe(15Å)/Mn(50Å)]_{40}(4)$. This calculation also shows a remarkable enhancement of the Bragg intensity at the wavelength just below the absorption edge as experimentally observed (Fig.2.2.3).

2.5 SYNCHROTRON RADIATION X-RAY DIFFRACTION STUDIES

Recently very brilliant synchrotron radiation (SR) x-rays have become available and have led to important advances in basic and applied science (25). The SR has a number of outstanding properties such as (a) high brilliance of source because of the small source size and the highly-collimated intense radiation, (b) a continous spectrum in energy and pulse duration in time and (c) a highly-polarized radiation. This section discusses how these special capabilities have been employed for the structural investigation and characterization of superlattices.

2.5.1 Anomalous scattering

The continuous spectrum of SR makes it possible to tune the x-ray energy at the absorption edge where the relevant atomic scattering factor changes significantly. Therefore a large increase in contrast of scattering-power can be obtained even for a superlattice consisting of two adjacent elements in the periodic table. Nakayama et al.(4) carried out such an experiment on an Fe/Mn superlattice, in which the scattering factor for the 1st-order reflection is enhanced roughly by a factor of $f_{Fe} - f_{Mn} = (f_{Fe}^0 - f_{Mn}^0) + [(\Delta f'_{Fe} - \Delta f'_{Mn}) + i(\Delta f''_{Fe} - \Delta f''_{Mn})] = 1.0 + [9.0 + 0.13i]$ at the Mn absorption edge ($\lambda_K = 1.896$ Å) as previously shown in Fig. 2.2.3(a). Such an enhancement of intensity was experimentally observed on $[Fe(15Å)/Mn(50Å)]_{40}$ as displayed in Fig.2.2.3(b). By tuning the x-ray wavelength at 1.900 Å, they successfully observed the small-angle Bragg reflections up to the 7th order as shown here in Fig.2.5.1. With a conventional Cu-Kα source (1.542 Å), on the other hand, only the 1st-order reflection was faintly seen because of the small contrast of the scattering power and a large background due to the fluorescence. These SR data for the Fe/Mn superlattice were well fitted by a trapezoidal model (Fig.2.3.1) with an interfacial thickness of about 10 Å. A similar anomalous scattering experiment

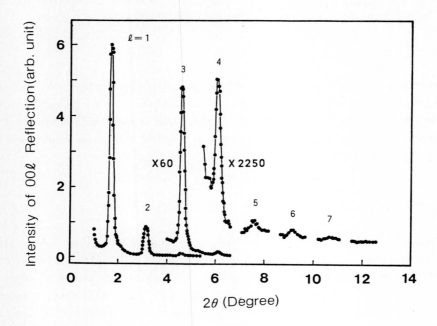

Fig. 2.5.1 Bragg reflections in the small-Q region observed on [Fe(15Å)/Mn(50Å)]$_{40}$ when the wavelength of synchrotron radiation x-rays was tuned at λ = 1.900Å slightly longer than the K absorption edge of Mn. The observed and calculated wavelength dependence of the 1st-order peak are shown in Figs.2.2.3(b) and 2.4.4, respectively. [From Nakayama et al.(4)]

was also made on Fe/V superlattices(26).

In contrast to the above example to determine the superlattice structure, Barbee et al.(27) used a superlattice with a well-known structure to determine the anomalous scattering factors. They made precise measurements of both reflectivity and the 1st-order Bragg reflection on a Ti/C superlattice in a wide energy range across the K-edge of Ti (E_K = 4.965 keV). Their data were fitted by calculations, in which the anomalous scattering factor $\Delta f'$ of Ti was parameterized. Prior to this fitting, $\Delta f''$ was obtained separately from their EXAFS (Extended X-ray Absorption Fine Structure) measurement of Ti. In this case, the superlattice is used as a gauge to determine physical quantities.

The EXAFS technique, which provides important information on the local environment around a target atom, is very useful for studying the interface structure of superlattices. To our knowledge, however, only a Nb/Zr superlattice(28) was investigated so far. The complementary use of EXAFS and diffraction data can lead to a more complete understanding of superlattice structures.

2.5.2 Magnetic scattering

Intense SR has made it possible to detect magnetic scattering which is weaker than the usual electron charge scattering (Thomson scattering) by a factor of $10^{-5} \sim 10^{-6}$. It has been demonstrated that magnetic x-ray scattering in corporation with the special capabilities of SR, such as high–momentum resolution and energy tunability, can provide a new insight into the magnetic structure and related phenomena which cannot be attained by magnetic neutron scattering. By using an interference effect between the charge and magnetic scattering, Vettier et al.(29) studied the modulation of the magnetic moment in an epitaxially–grown Gd/Y superlattice which had twenty–one $(00.1)_{hcp}$ planes in each layer. The diffraction intensity due to the interference is given as

$$ I \propto F'^2 + F''^2 - \frac{2\lambda_c}{\lambda} F''(\mathbf{k}_i \times \mathbf{k}_f) \cdot \mathbf{S} , \qquad (2.42) $$

where $\lambda_c = h/mc$ and $\mathbf{S}$ is the magnetic structure factor whose direction can be reversed by an external magnetic field. Therefore the flipping ratio, which is defined as the relative intensity change on field reversal, is calculated as

$$ R = (I_\uparrow - I_\downarrow)/(I_\uparrow + I_\downarrow) \propto F''S_\perp/(F'^2 + F''^2) , \qquad (2.43) $$

where $S_\perp$ is the projection of $\mathbf{S}$ on the normal to the scattering plane. As one can see in Eq.(2.43), the flipping ratio is sensitive to the imaginary part of the crystal structure factor which can be significantly varied by a change of x-ray energy across the absorption edge. Figure 2.5.2 shows their experimental set–up and the energy dependence of the flipping ratio observed for the 2nd and 3rd order satellites around the 400 reflection across the L_{II} and L_{III} edges of Gd. The solid curves are calculated values based on the assumption of a full Gd moment at the center with a smooth reduction in the projected moment as the interface is approached. It should be mentioned that in their analysis the strain modulation was carefully taken into account to separate it from the magnetic contribution.

In addition to this interference magnetic scattering, pure magnetic x-ray scattering study should be studied in the near future to reveal characteristics of the magnetic structure of superlattices as bulk Ho and Er were extensively investigated very recently(30).

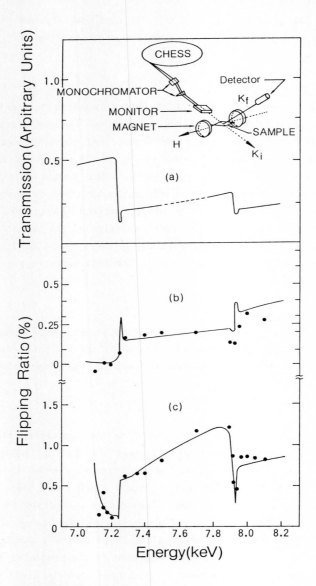

Fig. 2.5.2 (a) Transmission through a Gd foil as a function of x-ray energy near the $L_{II}(7.9310$ keV) and $L_{III}(7.2430$ keV) absorption edges.   As shown in the experimental set-up, the scattering plane is vertical while the magnetic field is applied **perpendicular to it.** The energy dependence of the flipping ratio observed at 200K (circles) and model-calculation (solid curves) are displayed for (b) the 2nd- and (c) the 3rd-order satellites around the 400 reflection of an epitaxially-grown Gd/ Y superlattice. [From Vettier et al.(29)]

2.5.3 Standing—wave method

It is well known that a standing—wave field is excited when x—rays are Bragg—reflected by a perfect crystal(5). As shown in Fig.2.5.3, the antinodal plane of the standing—wave with its wavelength identical to the relevant lattice spacing is located at the middle point of the adjacent diffracting planes when the glancing angle of the beam is slightly lower than the center of the Darwin curve. As the glancing angle becomes higher beyond the Bragg angle, its location changes continuously toward the diffracting plane position. Therefore, the yield of fluorescence x—rays or secondary electrons emitted from target atoms becomes maximum when the antinodal plane positions at them and its field is strong enough to excite such emissions. Thus by using this phenomenon, one can identify the sites of foreign atoms such as impurity and chemisorbed atoms.

In a process of making superlattices by the sputtering method, Ar^+ ions are entrapped as an impurity. Matsushita et al.(31) measured the yield of Ar—$K\alpha$ fluorescence x—rays (2.96 keV) emitted from a $[W(12Å)/Si(18Å)]_{48}$ superlattice as a function of the glancing angle (rocking curve) as shown by the thick curves in Fig.2.5.4. Their data analysis led to the conclusion that the Ar^+ density in the Si layer was as high as that in the W layer by a factor of 2∿4.

Since the standing—wave technique is sensitive enough to probe a small amount of target atoms and to determine their sites, it should also be powerful for precisely determining the interplanar separation between two kinds of atomic layers at the interface of a superlattice.

2.5.4 Optical devices

Synthetic superlattices are capable of functioning as novel optical elements because their lattice—spacing and scattering—power density can be tailored to match required specifications. Although various applications of superlattices have been reported at recent conferences(32), we have concentrated on the x—ray monochromator in this section. Natural crystals hardly have their lattice—spacing more than 10 Å, which is easily produced in superlattices. According to Bragg's law, the monochromatization of a longer wavelength of x—ray requires a larger lattice—spacing. Roughly speaking, the guidelines for making a high—quality monochromator are given by Eq.(2.34). A higher reflectivity is obtained for a larger value of A, i.e. a higher contrast in scattering—power between two kinds of layers and the larger number of stacking layers. For this purpose, the combination of W and C is the most widely used for x—ray monochromators. One more important requirement for a superlattice monochromator is the thermal stability of its superlattice structure. Particularly SR loads it with powerful heat radiations. A W/C superlattice is tough enough to withstand such a heat loading. As previously

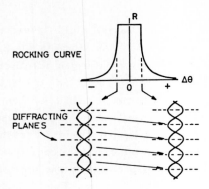

Fig. 2.5.3   X-ray standing-wave excited when Bragg-reflected in a perfect crystal.   As the   crystal is being rocked, its position changes with respect to the diffracting plane.

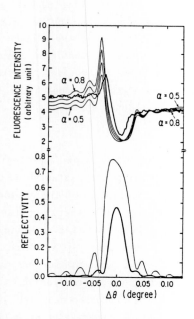

Fig. 2.5.4 Thick curves show the observed rocking curve of the 1st order Bragg reflection (bottom) and intensity of the Ar-Kα fluorescence x-rays emitted from the entrapped Ar ions (top) in $[W(12Å)/Si(18Å)]_{48}$ prepared by the sputtering method. Thin curves are the model-calculations for various values of α, which is a parameter representing the fact that the atomic density of Ar in Si layers is α/(1 - α) times higher than in W layers.
[From Matsushita et al.(31)]

mentioned in 2.3.2, Dinklage(22) first studied the structural stability of Au/Mg and Fe/Mg superlattices at elevated temperatures. Recently Fujii et al.(21) systematically measured the thermal annealing effect on Fe/Mg superlattices. Part of their data is shown in Fig.1.5.2 (Chapter 1).

Some x-ray experiments such as fluorescence analysis(33) and angle-dispersive EXAFS(34) require widely-monochromatized SR beams by an order of $\Delta E/E = 10^{-1} \sim 10^{-2}$. This requirement is not easily satisfied by conventional crystals which usually slice the energy band by an order of $\Delta E/E = 10^{-3}$. According to a differential form of Bragg's law given as $\Delta E \propto 2d\cos\theta \cdot \Delta\theta$, a larger energy band is obtained by a larger lattice-spacing d when the acceptance angle $\Delta\theta$ is fixed. The superlattice again satisfies this requirement and plays the role of a wide-band-pass monochromator(35).

Usually the quality of a superlattice monochromator is evaluated by measuring the reflectivity of the 1st-order Bragg reflection and its uniformity across the film plane. Ishikawa et al.(36) applied their special diffraction method to the characterization of W/C and W/Si superlattices deposited either on a silicon wafer or on an optical flat fused quartz. Their method, originally developed for characterizing x-ray mirrors(37), measures the angle-resolved x-ray scattering for both totally- and Bragg-reflected beams with extremely high angular resolution (about 4 seconds of arc). Their monochromator and analyzer, which play the role of a collimator, consist of quintuple-bounce (five-times consecutive reflections) Si 220 channel-cut crystals. Such extremely well collimated tailless beams make it possible to detect slope errors of the substrate very sensitively and serve as a critical test for optical devices.

The SR x-ray diffraction studies on superlattices started very recently. The number of such publications is not so many as yet but is growing very rapidly. Novel methods of SR x-ray diffraction should be developed to study the novel properties of superlattices.

2.6   SUMMARY

X-ray diffraction is a very powerful and nondestructive method for characterizing prepared superlattices and studying their structures. In this chapter, relatively simple methods have been introduced to analyze observed diffraction patterns with some assumptions. For example, only the one-dimensional model is used neglecting the full considerations of interface roughness and lateral ordering. These are still challenging problems which are not routinely analyzed. Also an interesting problem from the standpoint of diffraction phenomena is x-ray diffraction which takes place in an imperfect superlattice. The diffraction process must be treated differently because it lies between the kinematical and dynamical diffraction limits. Nevertheless, we hope that this chapter will help someone starting an x-ray diffraction experiment to understand the overall structural aspects of superlattices.

The recent availability of SR x-rays has led to important advances in understanding the structural as well as magnetic properties of superlattices and in applying superlattices as optical devices. Since the number of SR facilities is increasing rapidly and much brighter x-rays will become available, the use of SR x-ray diffraction will be significant in the study of superlattices in the near future.

I wish to express my sincere thanks to Y. Endoh, T. Ishihara, T. Ishikawa, K. Kawaguchi, C. F. Majkrzak, T. Matsushita, I. Moritani, N. Nakayama, T. Ohnishi, S. Sasaki, T. Shinjo and Y. Yamada for their fruitful discussion and/or collaboration.

REFERENCES

(1) D.B. McWhan, Synthetic Modulated Structures, edited by L.L. Chang and B.C. Giesen, Academic Press, 1985, 43pp.
(2) T.W. Barbee, Jr., Applications of Thin-Film Multilayered Structures to Figured X-Ray Optics, edited by G.F. Marshall, Proc. SPIE 563, 1985, 2pp.
(3) R. Pynn, Y. Fujii and G. Shirane, Acta Cryst. A39 (1983) 38.
(4) N. Nakayama, I. Moritani, T. Shinjo, Y. Fujii and S. Sasaki, in preparation.
(5) For example, W.H. Zachariasen, Theory of X-ray Diffraction in Crystals, Wiley, New York, 1945; B.W. Batterman and H. Cole, Rev. Mod. Phys. 36 (1964) 681.
(6) Y. Fujii, T. Ohnishi, T. Ishihara, Y. Yamada, K. Kawaguchi, N. Nakayama and T. Shinjo, J. Phys. Soc. Jpn. 55 (1986) 251.
(7) Y. Endoh, K. Kawaguchi, N. Hosoito, T. Shinjo, T. Takada, Y. Fujii and T. Ohnishi, J. Phys. Soc. Jpn. 53 (1984) 3481.
(8) P.F. Carcia and A. Suna, J. Appl. Phys. 54 (1983) 2000.
(9) M.R. Khan, C.S.L. Chun, G.P. Felcher, M. Grimsditch, A. Kueny, C.M. Falco and I.K. Shuller, Phys. Rev. B27 (1983) 7186.
(10) C.M. Falco, W.R. Bennett and A. Boufelfel, Dynamical Phenomena at Surfaces, Interfaces and Superlattices, edited by F. Nizzoli, K.-H. Rieder and R.F. Willis, Springer Verlag, 1985, 35pp.; L.H. Greene, W.L. Feldmann, J.M. Rowell, B. Batlogg, R. Hull and D.B. McWhan, in preprint.
(11) K. Kawaguchi, R. Yamamoto, N. Hosoito, T. Shinjo and T. Takada, J. Phys. Soc. Jpn. 55 (1986) 2375.
(12) T. Shinjo, N. Nakayama, I. Moritani and Y. Endoh, J. Phys. Soc. Jpn. 55 (1986) 2512.
(13) D.B. McWhan, M. Gurvitch, J.M. Rowell and L.R. Walker, J. Appl. Phys. 54 (1983) 3886.
(14) E.M. Gyorgy, D.B. McWhan, J.F. Dillon, Jr., L.R. Walker and J.V. Waszczak, Phys. Rev. B25 (1982) 6739.
(15) S.M. Durbin, J.E. Cunningham, M.E. Mochel and C.P. Flynn, J. Phys. F11 (1981) L223.
(16) J. Kwo, D.B. McWhan. M. Hong, E.M. Gyorgy, L.C. Feldman and J.E. Cunningham, Layered Structures, Epitaxy, and Interfaces, edited by J.H. Gibson and L.R. Dawson, Proc. Material Res. Soc. Symp. 37 (1985) 509.
(17) L. Esaki, L.L. Chang and R. Tsu, Proc. of the 12th International Conf. on Low Temp. Phys., edited by E. Kanda, 1971, 551pp.

(18) A. Segmüller and A.E. Blakeslee, J. Appl. Cryst. 6 (1973) 19.
(19) R.F. Miceli, H. Zabel and J.E. Cunningham, Phys. Rev. Lett. 54 (1985) 917.
(20) M. Jalochowski and P. Mikolajczak, J. Phys. F13 (1983) 1973.
(21) Y. Fujii, T. Ishihara, T. Ohnishi, N. Nakayama, K.Kawaguchi, I. Moritani, T. Shinjo and Y. Endoh, in preparation.
(22) J.B. Dinklage, J. Appl. Phys. 38 (1967) 3781.
(23) J.H. Underwood and T.W. Barbee, Jr., Appl. Optics 20 (1981) 3027.
(24) W.C. Marra, P. Eisenberger and A.Y. Cho, J. Appl. Phys. 50 (1979) 6927.
(25) For example, Synchrotron Radiation Research, edited by H. Winick and S. Doniach, Plenum Press, 1980; Handbook on Synchrotron Radiation, edited by D.E. Eastman and Y. Farge, North-Holland Pub.,1983.
(26) N. Nakayama, I. Moritani, T. Shinjo, Y. Fujii and S. Sasaki, private communications.
(27) T.W. Barbee, Jr., W.K. Warburton and J.H. Underwood, J. Opt. Soc. Am. B1 (1984) 691.
(28) T. Claeson, J.B. Boyce, W.P. Lowe and T.H. Geballe, Phys. Rev. B29 (1984) 4969.
(29) C. Vettier, D.B. McWhan, E.M. Gyorgy, J. Kwo, B.M. Buntschuh and B.W. Batterman, Phys. Rev. Lett. 56 (1986) 757.
(30) D. Gibbs, D.E. Moncton, K.L. D'Amico and B.H. Grier, Phys. Rev. Lett. 55 (1985) 234; D. Gibbs, J. Bohr, J.D. Axe, D.E. Moncton and K.L. D'Amico, to be published.
(31) T. Matsushita, A. Iida, T. Ishikawa, T. Nakagiri and K. Sakai, Nucl. Instrum. & Meth. A246 (1986) 751.
(32) Applications of Thin-Film Multilayered Structures to Figured X-Ray Optics, Proc. SPIE 563, 1985; Proc. of International Conf. on X-ray and VUV Synchrotron Radiation Instrum., edited by G.S. Brown and I. Lindau, Nucl. Instrum. & Meth. A246 (1986).
(33) A. Iida, T. Matsushita and Y. Goshi, Nucl. Instrum. & Meth. A235 (1985) 597.
(34) T. Matsushita and R.P. Phizackerley, Jpn. J. Appl. Phys. 20 (1981) 2223; R.P. Phizackerley, Z.U. Rek, G.B. Stephenson, S.D. Conradson, K.O. Hodgson, T. Matsushita and H. Ohyanagi, J. Appl. Cryst. 16 (1983) 220.
(35) T. Matsushita and H. Hashizume, Handbook on Synchrotron Radiation Vol.1a, edited by E.-E. Koch, North-Holland Pub., 1983, 261pp.
(36) T. Ishikawa, A. Iida and T. Matsushita, Nucl. Instrum. & Meth. A246 (1986) 348.
(37) T. Matsushita, T. Ishikawa and K. Kohra, J. Appl. Cryst. 17 (1984) 257.

TABLE OF CRYSTAL DATA AND X-RAY ENERGIES (EMISSION AND ABSORPTION) OF ELEMENTS

ATOMIC NUMBER		ATOMIC WEIGHT	X'TAL SYSTEM (phase)	LATTICE PARAMETERS (Å, deg)[1]	PRINCIPAL EMISSION LINES (keV)[2]					ABSORPTION EDGES (keV)[2]			
					$K\alpha_1$	$K\alpha_2$	$K\beta_1$	$L\alpha_1$		K	L_I	L_{II}	L_{III}
3	Li	6.94	bcc(β)	a=3.5100	0.0543				Li	0.0548			
4	Be	9.01	hcp(α)	a=2.286 c=3.584	0.1085				Be	0.111			
5	B	10.81	rhom(α)	a=5.057 α=58.07	0.1833				B				
6	C	12.01	dia	a=3.5670	0.277				C	0.2838			
7	N	14.01			0.3924				N	0.4000			
8	O	16.00			0.5249				O	0.5317			
9	F	19.00			0.6768				F				
10	Ne	20.18			0.8486				Ne	0.8669			
11	Na	22.99	bcc(β)	a=4.2906	1.0410		1.0711		Na	1.0717	0.0628		0.0306
12	Mg	24.31	hcp	a=3.2094 c=5.2105	1.2536		1.3022		Mg	1.3034	0.0870	0.0497	0.0495
13	Al	26.98	fcc	a=4.0496	1.4867	1.4863	1.5575		Al	1.5599			0.0728
14	Si	28.09	dia	a=5.4305	1.7400	1.7394	1.8359		Si	1.8400			0.1006
15	P	30.97	bcc	a=18.8	2.0137	2.0127	2.1390		P	2.1435			0.132
16	S	32.06	ortho	a=10.4646 b=12.8660 c=24.4860	2.3078	2.3066	2.4640		S	2.4705			
17	Cl	35.45			2.6224	2.6208	2.8156		Cl	2.8196			
18	Ar	39.95			2.9577	2.9556	3.1905		Ar	3.2029			
19	K	39.10	bcc	a=5.247 [78K]	3.3138	3.3111	3.5896		K	3.6078			0.2946
20	Ca	40.08	fcc(α)	a=5.5884	3.6917	3.6881	4.0127	0.3413	Ca	4.0381		0.3529	0.3493
21	Sc	44.96	hcp(α)	a=3.3090 c=5.2733	4.0906	4.0861	4.4605	0.3954	Sc	4.489			
22	Ti	47.90	hcp(α)	a=2.9511 c=4.6843	4.5108	4.5049	4.9318	0.4522	Ti	4.9645			0.4544
23	V	50.94	bcc	a=3.0231	4.9522	4.9446	5.4273	0.5113	V	5.4639	0.741	0.691	0.598
24	Cr	52.00	bcc	a=2.884	5.4147	5.4055	5.9467	0.5728	Cr	5.9888			

Z	El	At. wt.	Structure	Lattice const.					El				
25	Mn	54.94	bcc(α)	a=8.9139	5.8988	5.8877	6.4905	0.6374	Mn	6.5376			
26	Fe	55.85	bcc(α)	a=2.8664	6.4038	6.3908	7.0580	0.7050	Fe	7.1112		0.7208	0.7074
27	Co	58.93	hcp(α)	a=2.507, c=4.070	6.9303	6.9153	7.6494	0.7762	Co	7.7095		0.7938	0.7790
28	Ni	58.70	fcc	a=3.5238	7.4782	7.4609	8.2647	0.8515	Ni	8.3317		0.8706	0.8536
29	Cu	63.55	fcc	a=3.6147	8.0478	8.0278	8.9053	0.9297	Cu	8.9803		0.9527	0.9331
30	Zn	65.38	hcp	a=2.6649, c=4.9468	8.6389	8.6158	9.5720	1.0117	Zn	9.6607	0.9495	1.0452	1.0220
31	Ga	69.72	ortho	a=4.523, b=7.661, c=4.524	9.2517	9.2248	10.2642	1.0979	Ga	10.3682	1.3028	1.1450	1.1169
32	Ge	72.59	dia	a=5.6575	9.8864	9.8553	10.9821	1.1880	Ge	11.1036	1.4132	1.2494	1.2170
33	As	74.92	rhomb	a=4.1318, α=54.13	10.5437	10.5080	11.7262	1.2820	As	11.865	1.5293	1.3587	1.3235
34	Se	78.96	hex	a=4.3656, c=4.9590	11.2224	11.1814	12.4959	1.3791	Se	12.6545	1.6525	1.4747	1.4340
35	Br	79.90			11.9242	11.8776	13.2914	1.4804	Br	13.470	1.781	1.599	1.5530
36	Kr	83.80			12.649	12.598	14.112	1.5860	Kr	14.3244	1.915	1.7297	1.6772
37	Rb	85.47	bcc	a=5.70	13.3953	13.3358	14.9613	1.6941	Rb	15.2023	2.063	1.8661	1.8067
38	Sr	87.62	fcc(α)	a=6.0849	14.1650	14.0979	15.8357	1.8066	Sr	16.107	2.217	2.0085	1.9411
39	Y	88.91	hcp	a=3.6474, c=5.7306	14.9584	14.8829	16.7378	1.9226	Y	17.038	2.377	2.1540	2.0794
40	Zr	91.22	hcp(α)	a=3.2312, c=5.1477	15.7751	15.6909	17.6678	2.0424	Zr	17.9989	2.541	2.3053	2.2225
41	Nb	92.91	bcc	a=3.3066	16.6151	16.5210	18.6225	2.1659	Nb	18.9869	2.710	2.4641	2.3706
42	Mo	95.94	bcc	a=3.1468	17.4793	17.3743	19.6083	2.2932	Mo	20.0039	2.881	2.6274	2.5234
43	Tc	(98)			18.3671	18.2508	20.619	2.4240	Tc	21.0473	3.055	2.7948	2.6780
44	Ru	101.07	hcp	a=2.7058, c=4.2816	19.2792	19.1504	21.6568	2.5586	Ru	22.1193	3.233	2.9663	2.8377
45	Rh	102.91	fcc	a=3.8044	20.2161	20.0737	22.7236	2.6967	Rh	23.2198	3.417	3.1448	3.0021
46	Pd	106.4	fcc	a=3.8907	21.1771	21.0201	23.8187	2.8386	Pd	24.348	3.607	3.3303	3.1730
47	Ag	107.87	fcc	a=4.0862	22.1629	21.9903	24.9424	2.9843	Ag	25.5165	3.8072	3.5258	3.3510
48	Cd	112.41	hcp	a=2.9788, c=5.6167	23.1736	22.9841	26.0955	3.1337	Cd	26.7159	4.0190	3.7280	3.5376
49	In	114.82	fct	a=4.5979, c=4.9467	24.2097	24.0020	27.2759	3.2869	In	27.9420	4.2373	3.9393	3.7302
50	Sn	118.69	dia(α)	a=6.4892	25.2713	25.0440	28.4860	3.4440	Sn	29.1947	4.4648	4.1573	3.9288

(continued)

			structure	lattice	Kα₁	Kα₂	Kβ₁	Lα₁	K		L_I	L_II	L_III
51	Sb	121.75	rhom	a=4.5067 α=57.10	26.3591	26.1108	29.7256	3.6047	30.4860	Sb	4.6984	4.3819	4.1323
52	Te	127.60	hex	a=4.4566 c=5.9268	27.4723	27.2017	30.9957	3.7693	31.8114	Te	4.9397	4.6126	4.3418
53	I	126.90	ortho [110K]	a=7.136 b=4.686 c=9.784	28.6120	28.3172	32.2947	3.9377	33.1665	I	5.192	4.8540	4.5587
54	Xe	131.30	bcc	a=6.079 [173K]	29.779	29.458	33.624	4.1099	34.59	Xe	5.4528	5.1037	4.7822
55	Cs	132.91			30.9728	30.6251	34.9869	4.2865	35.987	Cs	5.721	5.3581	5.0113
56	Ba	137.34	bcc	a=5.013	32.1936	31.8171	36.3782	4.4663	37.452	Ba	5.996	5.6233	5.2470
57	La	138.91	hex(α)	a=3.770 c=12.159	33.4418	33.0341	37.8010	4.6510	38.934	La	6.268	5.889	5.484
58	Ce	140.12	fcc(γ)	a=5.1601	34.7197	34.2789	39.2573	4.8402	40.453	Ce	6.548	6.161	5.723
59	Pr	140.91	hex(α)	a=3.6725 c=11.8354	36.0263	35.5502	40.7482	5.0337	42.002	Pr	6.834	6.439	5.963
60	Nd	144.24	hex(α)	a=3.6579 c=11.7992	37.3610	36.8474	42.2713	5.2304	43.574	Nd	7.1294	6.7234	6.2092
61	Pm	(147)	hex	a=3.621 c=26.25	38.7247	38.1712	43.826	5.4325	45.198	Pm	7.436	7.014	6.4605
62	Sm	150.35			40.1181	39.5224	45.413	5.6361	46.849	Sm	7.7478	7.3132	6.7172
63	Eu	151.96	bcc	a=4.5820	41.5422	40.9019	47.0379	5.8457	48.519	Eu	8.0607	7.6199	6.9806
64	Gd	157.25	hcp(α)	a=3.6360 c=5.7826	42.9962	42.3089	48.697	6.0572	50.233	Gd	8.3864	7.9310	7.2430
65	Tb	158.92	hcp(α)	a=3.6010 c=5.6936	44.4816	43.7441	50.382	6.2728	52.002	Tb	8.7167	8.2527	7.5153
66	Dy	162.50	hcp(α)	a=3.5903 c=5.6475	45.9984	45.2078	52.119	6.4952	53.793	Dy	9.0548	8.5830	7.7897
67	Ho	164.93	hcp(α)	a=3.5773 c=5.6158	47.5467	46.6997	53.877	6.7198	55.619	Ho	9.3994	8.9164	8.0676
68	Er	167.26	hcp(α)	a=3.5588 c=5.5874	49.1277	48.2211	55.681	6.9487	57.487	Er	9.7574	9.2622	8.3575
69	Tm	168.93	hcp(α)	a=3.5375 c=5.5546	50.7416	49.7726	57.517	7.1799	59.38	Tm	10.1206	9.6171	8.6496
70	Yb	173.04	fcc	a=5.4862	52.3889	51.3540	59.37	7.4156	61.30	Yb	10.4904	9.9761	8.9441
71	Lu	174.97	hcp(α)	a=3.5031 c=5.5509	54.0698	52.9650	61.283	7.6555	63.31	Lu		10.3448	9.2490

No.	Symbol	At. wt.	Structure	Lattice								
72	Hf	178.49	hcp(α)	a=3.1946 c=5.0511	55.7902	54.6114	63.234	7.8990	65.31	11.274	10.7362	9.5577
73	Ta	180.95	bcc	a=3.298	57.532	56.277	65.223	8.1461	67.403	11.682	11.132	9.8766
74	W	183.85	bcc	a=3.1652	59.3182	57.9817	67.2443	8.3976	69.508	12.0996	11.538	10.1999
75	Re	186.21	hcp	a=2.760 c=4.458	61.1403	59.7179	69.310	8.6525	71.658	12.530	11.954	10.5306
76	Os	190.2	hcp	a=2.7353 c=4.3191	63.0005	61.4867	71.413	8.9117	73.856	12.972	12.381	10.8683
77	Ir	192.22	fcc	a=3.8389	64.8956	63.2867	73.5608	9.1751	76.101	13.423	12.820	11.212
78	Pt	195.09	fcc	a=3.9239	66.832	65.122	75.748	9.4423	78.381	13.883	13.2723	11.562
79	Au	196.97	fcc	a=4.0785	68.8037	66.9895	77.984	9.7133	80.720	14.3537	13.7361	11.9212
80	Hg	200.59	rhom [227K]	a=3.005 α=70.53	70.819	68.895	80.253	9.9888	83.109	14.842	14.215	12.286
81	Tl	204.37	hcp(α)	a=3.4566 c=5.5248	72.8715	70.8319	82.576	10.2685	85.533	15.343	14.699	12.660
82	Pb	207.19	fcc	a=4.9502	74.9694	72.8042	84.936	10.5515	88.005	15.855	15.2053	13.0406
83	Bi	208.98	rhom	a=4.736 α=57.23	77.1079	74.8148	87.343	10.8388	90.534	16.376	15.719	13.426
84	Po	(209)	sc(α)	a=3.345	79.290	76.862	89.80	11.1308				
85	At	(219)			81.52	78.95	92.30	11.4268				
86	Rn	(222)			83.78	81.07	94.87	11.7270				
87	Fr	(223)			86.10	83.23	97.47	12.0313				
88	Ra	226.03			88.47	85.43	100.13	12.3397				
89	Ac	227.03			90.884	87.67	102.85	12.6520				
90	Th	232.04	fcc(α)	a=5.0845	93.350	89.953	105.609	12.9687	109.646	19.236	18.486	15.444
91	Pa	231.04			95.868	92.287	108.427	13.2907		20.464	19.683	
92	U	238.03	ortho(α)	a=2.8537 b=5.8695 c=4.9548	98.439	94.665	111.300	13.6147	115.62	21.771	20.945	17.165

References

1) Mainly referred to W. B. Pearson, A Handbook of Lattice Spacings and Structures of Metals and Alloys (Vol. 2), Pergamon Press, 1967.
2) J. A. Bearden, Rev. Mod. Phys. 39 (1967) 78.

(continued)

Remarks

(1) The lattice parameters were measured at room temperature unless otherwise specified.
(2) X-ray energy E can be converted into the wavelength λ or wave-number k by the following relation:

$$E(keV) = 12.398/\lambda(\mathring{A}) = 1.9732 \ k(\mathring{A}^{-1}).$$

(3) The intensity ratio of the K emission lines is approximately given by $K\alpha_1 : K\alpha_2 : K\beta_1 = 100 : 50 : 15$. Therefore, the averaged energy of the unresolved lines $K\alpha_1$ and $K\alpha_2$ is obtained as

$$E(K\bar{\alpha}) = [100 \ E(K\alpha_1) + 50 \ E(K\alpha_2)]/150.$$

(4) The natural energy width (FWHM) of the emission line for typical elements is as follows:

	E(keV)	ΔE(keV)	ΔE/E
Cu $K\alpha_1$	8.0478	0.0030	3.8×10^{-4}
$K\alpha_2$	8.0278	0.0040	5.0×10^{-4}
$K\beta_1$	8.9053	0.0090	1.0×10^{-3}
Mo $K\alpha_1$	17.4793	0.0071	4.1×10^{-4}
$K\alpha_2$	17.3743	0.0078	4.5×10^{-4}

Notice that the unresolved $K\alpha_1$ and $K\alpha_2$ lines of Cu, for example, result in the effective energy broadening ΔE/E = 2.5×10^{-3}, which is as large as the natural width by an order of magnitude.

Chapter 3

Neutron Diffraction Studies
on Metallic Superlattices

Y. ENDOH and C. F. MAJKRZAK*
Tohoku University
*Brookhaven National Laboratory

3.1   INTRODUCTION

Considerable efforts have been made in recent years to describe the magnetic states of surfaces and interfaces(1,2). Magnetic layers made up of a discrete number of atomic planes of moments can be deposited alternately with non-magnetic layers of a given thickness as a model system for investigating:

a)   the effects of reduced dimensionality on the magnetic ground state or microscopic configuration of atomic moments in the magnetic layers;

b)   the temperature and field dependence of the magnetic arrangement, including critical phenomena;

c)   interlayer magnetic coupling;

and d)   the interaction of magnetic layers with superconducting layers.

Because neutrons scatter coherently from ordered arrays of atomic moments, long-range order on a microscopic scale can be probed. In the case of synthetic superlattices advantage can be taken of the artificial periodicity as in the case of x-ray diffraction studies of the chemical structures. Furthermore, the magnetic moment of the neutron couples directly to atomic moments with a scattering power that is comparable to that for the nuclear scattering which results from the other principal interaction of slow neutrons with condensed matter. For x-rays the magnetic interaction potential is about six orders of magnitude weaker than that which gives rise to charge scattering.

The role of neutron diffraction in the study of magnetic superlattices is described in this chapter. The fundamental principles of neutron diffraction pertinent to such studies are first discussed followed by a number of actual examples that are both physically interesting and illustrative of the power of the method. The practical instrumental uses of multilayers as polarizers and monochromators for neutrons (and x-rays) are described elsewhere (3-7).

3.2 POLARIZED NEUTRON DIFFRACTION

3.2.1 General theory

In the time since the pioneering work of Shull and Wollan (8) more than thirty-five years ago, neutron diffraction has become a well-established method for determining the microscopic chemical and magnetic structures of condensed matter and is now widely used by physicists, chemists and biologists (9). The methodology is well-documented and there exist a number of good, general texts on neutron scattering (see, for example, Refs. 10-19). To a large extent the formalism developed to understand the diffraction of x-rays by condensed matter can be applied to neutron diffraction. Much of the diffraction theory common to both x-rays and neutrons is presented in Chapter 2 of this book. In this chapter the essential differences will be addressed.

Given that the theoretical descriptions of neutron and x-ray diffraction are similar, the principal interaction of x-rays with matter is electronic whereas for neutrons two interactions predominate, one being nuclear and the other magnetic in nature. One important consequence is that neutrons are particularly sensitive to magnetic architectures. In fact, neutron scattering data constitute the most significant body of experimental evidence regarding long-range magnetic order in condensed matter. Insofar as the nuclear interaction is concerned, the strength varies erratically with isotope, though it is typically of the same order of magnitude, unlike the interaction of an x-ray photon with an atom which is proportional to the number of electrons and so increases regularly with atomic number. Neutrons can therefore be useful in distinguishing between atoms of neighboring atomic number or different isotopes of the same element and are sensitive to the positions of light elements, most notably hydrogen.

Although inelastic scattering processes are not of primary interest here, it is worthwhile noting that for the wavelengths of the order of the interatomic spacings in crystals which are required for diffraction studies, the corresponding neutron and x-ray energies differ greatly, the former being of the order of millivolts and the latter thousands of electron volts. Thermal neutron energies are therefore comparable to thermal fluctuations and magnetic excitations in solids and consequently the neutron is an ideal probe of the dynamics of condensed matter.

3.2.1.A Physical properties of the neutron

The neutron has a mass of 1.675×10^{-27} kg. slightly larger than that of the proton. A free neutron, in fact, decays into a proton accompanied by the emission of an electron and an antineutrino with a half-life of the order of a thousand seconds. The neutron is electrically neutral and is a Fermi particle with spin 1/2 and a magnetic dipole moment of -1.913 nuclear magnetons (the negative sign indicates that the magnetic moment is in the opposite direction to that of the spin). The neutron wavelength λ is related to its mass m and velocity v by the well-known quantum mechanical relation $\lambda = h/(mv)$, where h is Planck's constant. As mentioned above, the principle interactions of thermal or slow neutrons with matter are nuclear and magnetic. Other interaction potentials are several orders of magnitude weaker and are ordinarily of no consequence in conventional diffraction experiments treated in this text (20).

3.2.1.B Nuclear scattering

The nuclear scattering length for neutrons is in general different for different isotopes and, for a nucleus with spin I, there are two values of the scattering length, one corresponding to a parallel, the other an antiparallel alignment of neutron and nuclear spins. As mentioned earlier, the nuclear scattering length has no regular relationship to nuclear mass or charge. This is a result of the resonance scattering which occurs in addition to the "hard sphere" scattering (see, for example, Ref.21). Furthermore, the nuclear scattering length is in general complex. The imaginary part is directly related to the absorption cross section and the real part can be either positive or negative. Absorption is most often negligible (exceptions include certain isotopes of Cd, B, and Li and a number of rare earths such as Gd). In most cases the real part of the scattering length is practically constant over the thermal or slow neutron energy range, whereas the absorption cross section is proportional to the wavelength (except in the neighborhood of a resonance). Table 3.1 lists selected values of neutron scattering lengths b and absorption cross sections together with corresponding x-ray values. More extensive tables can be found in References 10 and 22.

For a rigidly bound nucleus, the scattering of neutrons is an elastic process and although the neutron-nucleus interaction is relatively strong, it is of sufficiently short range compared to thermal or slow neutron wavelengths (of the order of 0.1Å or greater) that the scattering is practically isotropic. This is different than the case for the scattering of x-rays from the electron distribution of an atom which is considerably more extended in space than the volume occupied by the atomic nucleus. For x-rays the scattering length $f=f(\theta)$ is not a constant but is dependent upon the scattering angle 2θ.

Element	Atomic Number	Isotope	Neutrons Reb (10^{-12}cm)	σ_{ABS} (10^{-24}cm^2)	X-rays Ref sinθ/λ=0 (10^{-12}cm)	sinθ/λ=0.5	σ_{ABS} (10^{-24}cm^2)
H	1	^{1}H	-0.37423	0.3326	0.28	0.02	0.7
		^{2}H	0.6674	0.000519	0.28	0.02	0.9
C	6	^{12}C	0.66535	0.00353	1.69	0.48	92.
N	7	^{14}N	0.937	1.91	1.97	0.53	175.
O	8	^{16}O	0.5805	0.00010	2.25	0.62	305.
Al	13	^{27}Al	0.3449	0.231	3.65	1.55	2180.
Si	14	*	0.4149	0.171	3.95	1.72	2830.
V	23	^{51}V	-0.04024	4.9	6.5	2.8	19700.
Fe	26	*	0.954	2.56	7.3	3.3	28600.
Ni	28	^{58}Ni	1.44	4.6	7.9	3.6	4450.
		^{62}Ni	-0.87	14.5	7.9	3.6	
Nb	41	^{93}Nb	0.7054	1.15	11.5	5.7	23600.
Cd	48	*	0.51	2520.	13.6	6.9	43000.
Pb	82	*	0.94003	0.171	23.1	12.9	79700.

Table 3.1   Selected values of the real part of the nuclear scattering lengths b and absorption cross sections σ_{ABS} for neutrons together with corresponding x-ray values (at λ =1.8Å for neutrons and λ = 1.54Å for x-rays). The asterisk* indicates values for the naturally occurring isotopic mixture. More extensive tables can be found in Ref.10, from which the values here were taken.

3.2.1.C Magnetic scattering

The magnitude of the magnetic scattering length p for neutrons is given by

$$p= \left[\frac{e^2\gamma}{2mc^2} \right] gJf= (0.27 \times 10^{-12} \text{ cm}) \, gJf \qquad (3.1)$$

where e, m and c are the electron charge, electron mass and speed of light, respectively, γ = 1.9 nuclear magnetons, g is the Lande splitting factor, and the quantum number J corresponds to the total angular momentum of the atom. The quantity f is an angle-dependent form factor analogous to the electronic form factor for x-rays. For neutrons f corresponds, however, to the Fourier transform of the unpaired electron density which gives rise to a net magnetic moment and which extends over a volume of space with linear dimensions comparable to the neutron wavelength, unlike the nucleus which is effectively a point scatterer.

Table 3.2 is a representative compilation of magnetic scattering length magnitudes. Thus, in order to accurately measure the structure factor corresponding to a magnetic crystal, it is necessary to know the angular dependence of the form factor.

$$p \ (10^{-12} cm)$$

Ion	$\frac{\sin \theta}{\lambda} = 0$	$\frac{\sin \theta}{\lambda} = 0.25 Å^{-1}$
Cr^{2+}	1.08	0.45
Mn^{2+}	1.35	0.57
Fe(metal)	0.60	0.35
Fe^{2+}	1.08	0.45
Fe^{3+}	1.35	0.57
Co(metal)	0.47	0.27
Co^{2+}	1.21	0.51
Ni(metal)	0.16	0.10
Ni^{2+}	0.54	0.23

Table 3.2   *Representative compilation of magnetic scattering length magnitudes.*

(After Table 16 of Ref.10).

On the other hand, by accurately measuring the integrated Bragg intensities for a single crystal with a known structure, it is in principle possible to deduce the magnetization distribution within the unit cell. In general, the more extended in space a component of the unpaired electron distribution is, the less it will contribute to the intensity of magnetically scattered neutrons at higher values of Q or scattering angle. It should also be noted that for some atoms, e.g., most of the rare earths, the magnetic moment arises from both spin and orbital momenta of the electrons and consequently the form factors are actually a combination of two components although only the resultant can be measured experimentally.

The interaction between the magnetic moments of neutron and atom depends not only on the magnitude of the moments but also on their orientation relative to one another and the scattering vector $\mathbf{Q}$, where $|\mathbf{Q}| = |\mathbf{k}_f - \mathbf{k}_i| = 2|\mathbf{k}_i|\sin \theta$ is the magnitude of the scattering vector perpendicular to the planes, $\mathbf{k}_f$ and $\mathbf{k}_i$ are the final and incident wavevectors, respectively , and θ is the Bragg angle. In order to take into account the vectorial nature of the interaction, it is useful to define a vector $\mathbf{q}$ so that

$$\mathbf{q} = \hat{\mathbf{Q}}(\hat{\mathbf{Q}}.\hat{\mathbf{S}}) - \hat{\mathbf{S}} \qquad\qquad (3.2)$$

where $\hat{Q}$ and $\hat{S}$ are unit vectors along the directions of the scattering vector and atomic magnetic moment respectively. The scattering amplitude U_j of the atom can then be described using the Pauli spin matrices (11,23) as

$$U_j = (+' \ -') \ [\ b_j + p_j \ \mathbf{q_j} \cdot \boldsymbol{\sigma} \ ] \ (\ {\textstyle + \atop -} \ ) \tag{3.3}$$

where b_j is the nuclear coherent scattering length of the jth atom, σ is the Pauli spin operator, and (+ -) and (+' -') refer to the spin eigenfunctions of the incident and scattered neutron, respectively [+ (-) implies spin parallel (antiparallel) to the quantization axis]. The equation above applies to a single isotope and a nucleus with zero spin but it is straightforward to generalize to the case of a nucleus with spin (see, for example, Ref.10). Expanding Eq.(3.3) with the specific stipulation that the z-axis be the direction of the neutron polarization, we obtain four scattering amplitudes, each pertaining to a particular spin-dependent process :

$$U_j^{++} = b_j + p_j \ q_{zj}$$

$$U_j^{--} = b_j - p_j \ q_{zj}$$

$$U_j^{+-} = p_j(q_{xj} + iq_{yj})$$

$$U_j^{-+} = p_j(q_{xj} - iq_{yj}) \tag{3.4}$$

The corresponding structure factors (introduced in Chapter 2 for x-rays) are then

$$F^{++}_{--} = \sum_{j=1}^{N} (b_j \ \pm \ p_j q_{zj} \ )e^{i\mathbf{Q} \ \mathbf{r}_j} \tag{3.5a}$$

$$F^{+-}_{-+} = \sum_{j=1}^{N} p_j(q_{xj} \ \pm \ iq_{yj} \ )e^{i\mathbf{Q} \ \mathbf{r}_j} \tag{3.5b}$$

where $\mathbf{r}_j$ is the position of the jth atom in the unit cell. Consider now some general as well as several important special consequences of the spin-dependent expressions for the structure factors.

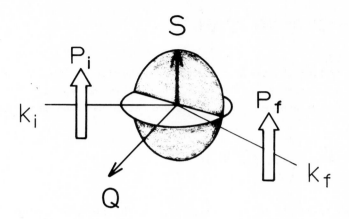

Fig. 3.2.1   Schematic drawing showing the relation among the
neutron wave vector **k**, the scattering vector **Q**, the magnetic
moment **S** and the neutron polarization **P** for one possible
configuration.   The subscript i or f indicates the initial or
final neutron state respectively.

Note first that coherent nuclear scattering is always non-spin-flip (NSF).
It can also be shown that nuclear spin-incoherent scattering is 2/3 spin-
flip(SF) and 1/3 NSF for unaligned nuclear moments regardless of the relative
orientation of the neutron polarization and scattering vector. Isotopic
incoherent scattering, on the other hand, is always NSF.

If a sample is ferromagnetic, a magnetic field of sufficient strength to
align individual domains must be applied to prevent depolarization of the
neutron beam. If magnetic moments are aligned along the direction of the
scattering vector, no magnetic scattering will occur since it is only the
projection of the magnetic moment in the reflecting plane (perpendicular to **Q**)
which contributes to the scattering. If, however, parallel moments are all
aligned perpendicular to **Q**, then there will be interference between the nuclear
and magnetic scattering so that $U_j^{\pm\pm} = b_j \pm p_j$ and $U_j^{\pm\mp} = 0$. By selecting $b_j = p_j$
in a monochromating crystal, only one spin state will be reflected. This is one
of the principal means by which a polarized beam can be obtained.

If, on the other hand, a sample is antiferromagnetic and **Q** is parallel to
the neutron polarization axis (**z** – axis), all magnetic scattering will be SF.

The advantage of using polarized neutrons in separating nuclear scattering is clearly demonstrated in the paper by Moon et al.(24).

3.2.2 Diffraction from superlattices

3.2.2.A Low angles

For diffraction at low Q, the scattering density profile of a bilayer can be treated as a continuum, whether the layers be epitaxial, textured, polycrystalline, or amorphous. The structure factor can then be written as

$$F = \int_{-\Lambda/2}^{\Lambda/2} \rho(z) \, e^{iQz} \, dz \qquad (3.6)$$

where Λ is the thickness of the bilayer, z is the distance measured along the normal to the plane of the film, and $\rho(z)$ is a three-dimensional scattering density which is in general spin-dependent. $\rho(z)$ is simply the product of the atomic density and scattering length at position z . If the number of coherent bilayers is sufficiently large that most of the diffracted radiation occurs as well-defined peaks at integer multiples of the bilayer periodicity, that is, at $Q_m = m2\pi/\Lambda$, then the intensity I_m of the mth reflection centered at Q_m will be proportional to $|F_m|^2$. It is then in principle possible to apply Fourier techniques to the problem of determining the scattering density of a multilayer. If, for example, a bilayer unit cell of length Λ is chosen so that $\rho(z)$ is an even function, then the scattering density can be written as a Fourier cosine series

$$\rho(z) = A_o + 2 \sum_{j=1}^{\infty} A_j \cos\theta \frac{2\pi jz}{\Lambda} \qquad (3.7)$$

where the A_j are the Fourier coefficients. It can be shown that the structure factor F_m evaluated at the low angle superlattice peak position given by $Q_m = m2\pi/\Lambda$ is proportional to the mth coefficient of the Fourier series expansion. Nevertheless, Fourier analysis can be limited in practice by an insufficient number of observable higher-order reflections and the uncertainty in the signs of the coefficients. It is then more practical to fit models of the scattering density profile to the diffracted intensity data directly. In fact, earlier studies on magnetic superlattices by polarized neutron diffraction were mostly done in this manner due to the insufficient number of observed reflections even at low angles. If, however, the number of coherent bilayers is so large that extinction effects become significant, then the dynamical diffraction theory must be applied (see Chapter 2).

It is sometimes necessary to account for mirror reflection and refraction effects at small Q (see, for example, Ref.25). The proper method of analysis is then analogous to that used in the solution of thin film optical interference problems and amounts to solving the Schrödinger equation for a plane wave incident upon and propagating through a layered but continuous medium (26). Boundary conditions are imposed at each interface and the reflection and transmission coefficients subsequently evaluated as a function of Q. Random and systematic variations in bilayer thickness as well as absorption can be readily incorporated. This method is widely used in the design of multilayer monochromators and supermirrors(27-30).

In a case where the number of bilayers M is relatively small, M-2 subsidiary maxima can be observed between adjacent primary maxima at $Q_m = m2\pi/\Lambda$, if the instrumental resolution is good enough (31; see also Chapter 2). In this case it is more appropriate to compare the observed and calculated diffracted intensity profiles (i.e., intensity as a continuous function of Q over an extended range).

3.2.2.B High angles

If the superlattice consists of single crystal bilayers with an atomic plane spacing d along the normal to the film plane, then diffraction maxima will occur at higher angles as well, at integer multiples of $2\pi/d$. If the sets of atomic planes comprising each bilayer are coherent with one another, and the number of bilayers large enough, well-defined satellite reflections will occur about $Q = 2\pi/d$ at $Q_m = 2\pi/d \pm m2\pi/\Lambda$. If, on the other hand, the sets of N coherent atomic planes are not coherent relative to one another, then a diffraction profile characteristic of a finite number of planes will be centered at $2\pi/d$ (31; see again Chapter 2). The Fourier method can also be applied to the analysis of satellite intensities at high angles (32).

3.2.2.C Ferromagnetic multilayers

One of the most interesting applications of polarized neutron diffraction to the study of magnetic superlattices is the determination of the magnetization profile across the thickness of a thin ferromagnetic layer. In the case of a single crystal superlattice, it is in fact possible, in principle, to determine the magnetization of each of the N atomic planes within a layer. If the magnetization profile is then measured as a function of temperature, the critical exponents for the magnetization of each atomic plane, including that at the interface, can be deduced and the effects of reduced dimensionality and interface anisotropy inferred. Of course it is necessary that the microscopic chemical structure of the multilayer be known, since the magnetic properties depend upon crystallographic orientation and interdiffusion and are affected by the coherency strains that can arise from lattice mismatch (33). Furthermore, in certain cases the range of the magnetic interactions is of the same order as the thickness of the bilayer so that coupling between different ferromagnetic layers can occur. In fact, a possible crossover from three- to two-dimensional behavior in sufficiently thin and well-separated magnetic layers can be investigated. Nevertheless, if the chemical composition and strain modulations in the superlattice are determined (by x-ray diffraction and/or neutron diffraction in an applied field of sufficient strength to align the moments parallel to Q), then the magnetic scattering lengths and thereby the magnitudes of the magnetic moments in each atomic plane can be obtained provided that a sufficient number of satellite reflections are observed (see Eq. 3.5a and the subsequent discussion pertaining to the case of a polarized beam incident upon a ferromagnet).

3.2.3 Experimental methods

3.2.3.A Neutron sources

There are two principal sources of neutrons, the nuclear reactor and the spallation source. In the former case the high energy neutrons that are produced in the fissioning of uranium atoms within the reactor core are subsequently moderated by inelastic collisions with a moderating material such as heavy or light water. The continuous spectrum of neutron energies emerging from a beam hole is Maxwellian. For a typical moderator temperature of 300K, the spectrum peaks at a wavelength of approximately 1.13Å. Useful intensities are generally obtained from approximately 0.5 to 5Å.   With secondary "hot" or "cold" moderators, this wavelength range can be extended considerably. For example, with a liquid hydrogen moderator at 20K, the Maxwellian peak can be shifted to longer wavelengths to give a usable flux of neutrons with wavelengths of 20Å or more.

Spallation   sources, on the other hand, are pulsed in nature, providing bursts of neutrons at regular intervals. These accelerator based devices produce high-energy neutrons when a pulsed proton beam, for example, impinges on a target such as uranium. The high-energy neutrons which result from the nuclear reaction are then moderated as in the nuclear reactor source. However, the inherent time dependence on the neutron spectrum allows for energy analysis to be performed by time-of-flight (TOF) techniques. The following two schematic diffraction diagrams illustrate the two different experimental techniques corresponding to the two types of neutron sources.

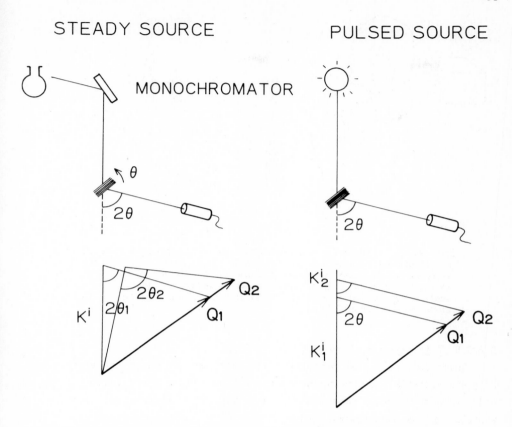

STEADY SOURCE

PULSED SOURCE

MONOCHROMATOR

Fig. 3.2.2   Schematic drawing of the diffraction geometry for two
types of neutron sources.   The upper figures are shown in   real
space and the bottoms in reciprocal space.

The steady flux of thermal neutrons from the reactor is monochromatized and
polarized with large crystals of ferromagnetic alloys such as Cu_2MnAl.   The
sample material is then exposed to the monochromatic and polarized neutrons,
and the diffraction patterns are obtained by performing $\theta - 2\theta$ scans as is
done in conventional x-ray diffraction techniques (Chapter 2).   However,
pulsed neutrons often use a wide wavelength range of spectrum simultaneously and
the wavelength or energy is determined by the TOF method.   Therefore the
diffraction pattern is given in the TOF spectrum of scattered neutrons at a
fixed diffraction angle once the energy spectrum of incoming neutrons to the
sample is determined.   The details of the neutron TOF diffraction technique
will not be discussed here and the interested reader is referred to the text by
Windsor (34).

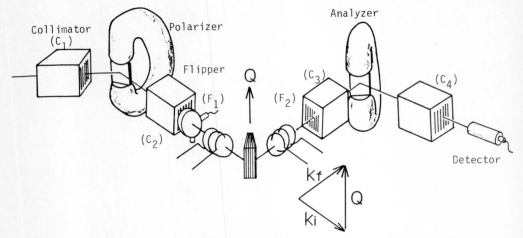

Fig. 3.2.3 *Schematic representation of a triple-axis spectrometer as described in detail in the text. The magnetic guide field is applied along the neutron flight path which is necessary to maintain an axis of quantization for a polarized neutron beam. The scattering vector Q lies in a horizontal plane defined by neutron wave vectors ki and kf.*

3.2.3.B Triple-axis spectrometer

Space does not allow for complete descriptions of all of the various types of neutron diffractometers currently in use. Instead, a triple-axis spectrometer will be described in some detail since most of the essential instrumental aspects can be illustrated, particularly in the more general case involving polarized beams.

Let us consider the main components of a triple-axis spectrometer, depicted schematically in Fig.3.2.3, sequentially, beginning with the monochromator. The monochromator is a single crystal which selects out of a collimated Maxwellian distribution of neutron wavelengths a nearly monochromatic beam by the Bragg diffraction process. Because of the finite angular divergence of the beam and the mosaic or natural width of the reflection, a relatively narrow but non-zero energy bandwidth is reflected. One of the most efficient non-polarizing monochromators available is pyrolytic graphite due to the high reflectivity. Other important monochromators include Cu, Ge and Be crystals. The properties of crystal monochromators for neutrons are discussed at length in an article by Freund and Forsyth (35). In order to obtain a polarized monochromatic beam a magnetic crystal such as a Heusler alloy, Cu_2MnAl, is used which has a nearly vanishing structure factor for one spin state. Thin-film multibilayer devices can also be used for polarized neutrons and are reviewed in Ref.3.

One problem with crystal monochromators is that the higher-order wavelengths $\lambda/2$, $\lambda/3$, etc. are also reflected in general, thereby contaminating the beam incident on the sample. These higher-order components can be suppressed by using a polycrystalline Be filter for primary wavelengths longer than 4Å or pyrolytic graphite filters which have several convenient primary wavelength "windows" in the thermal energy range.

Additional collimators between monochromator and sample, sample and analyzer, and analyzer and detector further define the angular divergences of incident and scattered beams and ultimately affect the overall instrumental resolution.

Neutrons scattered by the sample with the proper k_f are subsequently reflected by the analyzer crystal into the detector. Because coherent elastic scattering cross sections are usually orders of magnitude larger than those for inelastic processes, an analyzer is often unnecessary. However, in order to distinguish the four possible spin-dependent cross sections for polarized neutrons(discussed in Section 3.2.1.C), the polarizing monochromator and analyzer crystals are required as well as magnetic guide fields over the neutron flight path to maintain an axis of quantization for the neutron spin. In addition, adiabatic neutron spin flipping coils before and after the sample allow for selecting the incident spin state and for determining that of a scattered neutron.

The detection of radiation most often relies on some ionization process which can be amplified to produce a measurable electrical impulse. Because neutrons have no charge and the energies used in scattering studies are from 0.001 to 1eV, the neutron interaction with matter will normally not produce charged particles. The detection of neutrons has, therefore, to a large extent, been based on nuclear capture reactions which produce electrons, protons, α -particles and other heavier fission fragments.

3.2.3.C Instrumental resolution

The monochromatic beam of neutrons incident upon the sample is in actuality composed of a distribution of magnitudes and directions of wave vectors k_i. The instrumental resolution depends on: 1) the interatomic plane spacing for a particular Bragg reflection from monochromator, sample and analyzer; 2) the corresponding widths of the angular distributions of microcrystallites (for mosaic single crystals); 3) the angular divergences of the incident and scattered beams as defined by the collimators; and 4) the relative orientation of the scattering vectors of monochromator, sample and analyzer. The resolution properties (e.g., energy width, widths in reciprocal space parallel and perpendicular to the mean scattering vector Q_0, etc.) corresponding to a particular spectrometer configuration can, in general, be calculated in a straightforward manner (36). A typical resolution of the order of 0.5 meV Full Width at Half Maximum (FWHM) in energy and 0.015 Å^{-1} (FWHM) parallel to Q_0 can be obtained over a significant range of sample scattering angles at λ = 2.35Å using a pyrolytic graphite monochromator and analyzer ((0002) reflection and mosaics of 25 minutes FWHM) with 20 minute horizontal collimations in the "W" configuration depicted in Fig.3.2.3.

3.3 ILLUSTRATIVE EXAMPLES

3.3.1 Early studies

Early neutron diffraction work on multilayer systems has been summarized in review articles by Endoh (37,38). In many cases, due to either a lack of detailed characterization of the chemical microstructure of the films or an insufficient number of observable higher-order reflections (because, for example, of cumulative errors in the superlattice period), only qualitative or semiquantitative   conclusions could be drawn from analyses of the neutron diffraction data alone.

Perfect lattice matching and interdiffusion are sometimes two contradictory factors: At the elevated temperatures during sample preparation, a close lattice matching between each constituent metal, or the structural coherency over the layers of two different compositions is substantially achieved, but, on the other hand, interdiffusion through the interface often occurs simultaneously. Magnetic superstructure films consisting of 3d transition metals are not ideal. It is still difficult to prepare highly oriented as well as well-ordered materials, although certain combinations such as Fe vs. V or Fe vs. Mg limit interdiffusion to within a few atomic layers.   In spite of these important limitations on the sample preparations, an attempt was made to determine the magnetization density distribution over the magnetic layers by polarized neutron diffraction. The unidirectional magnetic density distribution in Fe layers in Fe/Sb superstructure films was determined from the flipping ratios of the first order magnetic reflection using Fe/Sb films of different compositions. In other words, the thickness of the Fe layers was varied with respect to the bilayer thickness, which was kept constant. The idea is that the flipping ratios are sensitive to the magnetic density distribution, not to the composition.   Then the results for the flipping ratios as a function of the composition were analyzed by comparison to model magnetic density distributions. The observed flipping ratios were described by a model in which the Fe layer just overlayering on the Sb layers becomes magnetically dead.   This interesting picture was proposed by Shinjo et al. by Mössbauer measurement of the selective deposition of ^{57}Fe isotope onto Fe layers.   The results indicate the formation of a new non-magnetic compound at the interface when Fe covered the Sb layers. A detailed description will be found in the following chapter.

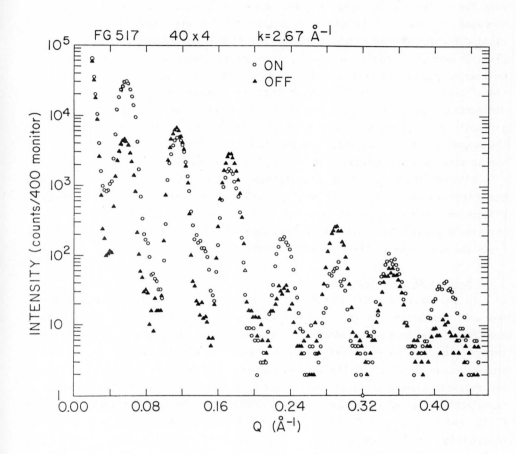

Fig. 3.3.1   Polarized neutron diffraction data at low Q for an
Fe/Ge multilayer. Seven diffraction peaks centered at values of
$Q=m2\pi/\Lambda$ are shown. The "ON"("OFF") data points correspond to
incident neutrons in the "+" ("−") spin eigenstate. The fact that
the flipping ratio is not greater than unity for all orders implies
a nonuniform magnetization in the ferromagnetic layer along the
modulation direction(39).

Interdiffusion is also a limiting factor in the study of the magnetic properties of many multilayer systems. Figure 3.3.1 shows polarized neutron data for an Fe/Ge multilayer at small values of Q (39). The Fe/Ge multilayer is composed of layers of Fe microcrystallites, which are highly oriented in the [110] direction normal to the plane of the film (but randomly oriented in the plane), and of amorphous Ge layers. In addition, there is an FeGe alloy region between adjacent Fe and Ge layers which is believed to account for the significant reduction in Fe moment that is deduced from an analysis of the diffraction data (39). Unfortunately, if interdiffusion occurs, it becomes difficult to distinguish the effects of alloying and reduced dimensionality on the magnetization at the interface even though the actual magnetization profile can be accurately determined from the diffraction measurements.

Although there have been many studies on superlattices of transition metals, quantitative conclusions concerning interesting issues on the interfacial effects on the magnetism still remain. Technical difficulties have arisen in the sample preparation, but the current pace of technological development should close the gaps between theories and experiments.

3.3.2 Rare earth superlattices

In the case of metallic, magnetic rare earths (RE), the magnetic moments are well localized and the indirect exchange interaction is via conduction electrons. The long-range nature of this Ruderman-Kittel-Kasuya-Yoshida (RKKY) interaction can be expected to give rise to a modulation of the magnetic properties in an artificially layered magnetic RE/non-magnetic RE structure. It has recently been discovered how to grow coherent, single crystal rare earth superlattices by molecular beam epitaxy (MBE) (40). RE superlattices of Gd and Y(41) and of Dy and Y (42) having relatively sharp interfaces (with interdiffusion limited to only a few atomic planes) and a high degree of uniformity in bilayer thickness have been produced. Neutron diffraction studies of the magnetic structures of these two systems are described below.

3.3.2.A Gd/Y

In a polarized neutron diffraction study (43) of $[Gd(NL)/Y(NL)]_M$ superlattices composed of M successive bilayers of NL basal planes of hexagonal-close-packed (hcp) Gd followed by NL such planes of Y, it was found that below the Curie temperature and in low fields the ferromagnetic Gd layers tend to align antiferromagnetically relative to one another for $NL_{Gd} = NL_Y = 10L$ in a microscopic antiphase domain structure that is coherent over many superlattice periods. For $NL_Y = 6L$ or $20L$, however, a simple long-range ferromagnetic order is observed. These results are consistent with a simple theoretical calculation which indicates that the RKKY coupling between Gd layers through the nonmagnetic

Y layers can be either ferromagnetic or antiferromagnetic in sign, depending on the Y layer thickness, and of sufficient strength to produce a long-range order.

Three samples, $[Gd(10L)/Y(10L)]_{225}$, $[Gd(10L)/Y(6L)]_{189}$ and $[Gd(10L)/Y(20L)]_{100}$, were studied using the spectrometer configuration shown schematically in the lower part of Fig.3.3.2. Neutrons are first simultaneously monochromated and polarized, then scattered by the sample (a flat coil flipper preceding the sample permits selection of either of the two neutron spin states). The second polarizing crystal or analyzer (along with another flipper) distinguishes the spin-flip (SF) from the non-spin-flip (NSF) scattering. A vertical guide field of the order of 10 to 20 Oe is maintained along the neutron flight path in order to define an axis of quantization for the neutron spin. With this arrangement four spin-dependent scattered intensities, $I_m^{++}(Q)$, I_m^{--} (Q), $I_m^{+-}(Q)$ and $I_m^{-+}(Q)$, can be measured as described above (see Section 3.2.1.C). The two superscripts denote the initial and final neutron spin states, respectively, and the subscript indicates the order of a satellite about a given primary Bragg reflection. Once again $Q = k_f - k_i$ where k_f and k_i are the final and initial neutron wavevectors, respectively. Satellite intensities were measured about the (0002) and (0004) reflections as well as the forward direction with Q parallel to the c-axis. After deconvolution from the instrumental resolution width, the natural widths of all of the observed satellite reflections were found to be approximately equal and to correspond to a coherence length of at least 1000Å or more than 17 superlattice periods.

Let us first discuss the $[Gd(10L)/Y(10L)]_{225}$ sample in detail, for which representative data taken at 150K in a field of 150 Oe is shown in Fig.3.3.2. The positions Q_m of the even-numbered satellites measured from the primary (002) reflection correspond to integer multiples of $2\pi/\Lambda_{SL}$, where $\Lambda_{SL} = 58.45$Å is the chemical modulation wavelength or bilayer thickness (i.e., $Q_m=(m/2) \cdot 2\pi/\Lambda_{SL}$). Above T_C (=290K) the intensities of the even-numbered satellites arise solely from the nuclear scattering associated with the superlattice chemical modulation so that $I_m^{++} = I_m^{--}$, whereas below T_C it is found that $I_m^{++} \neq I_m^{--}$ which indicates the presence of an additional ferromagnetic component. To within experimental accuracy, no spin-flip scattering is found to occur at the even-numbered satellite positions. The odd-numbered satellites, on the other hand, begin to appear below Tc in low fields at integer multiples of the wavevector for a doubled bilayer thickness (e.g., for m = 1, $Q_1 = (1/2) 2\pi/\Lambda_{SL}$, and for m = 3, $Q_3 = (3/2) 2\pi/\Lambda_{SL}$) with intensities corresponding to spin-flip scattering only. At 79K a field of several thousand Oe simultaneously saturates the magnetization and reduces the intensity of the odd-numbered satellites to within background.

Appropriate corrections were made for the instrumental polarizing and flipping efficiencies and a contribution to the superlattice(002) intensity due

98

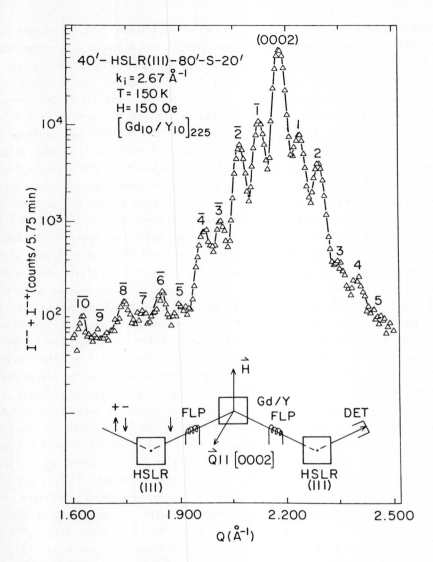

Fig. 3.3.2   Representative neutron diffraction data for a Gd/Y
superlattice with $N_{Gd} = N_Y = 10$ (the spectrometer configuration
including beam collimations is given in the upper left). Satellites
occur at integer multiples of half the superlattice wavevector with
odd- and even-numbered satellites corresponding to spin-flip and
non-spin-flip scattering respectively , as described in the text.
The lower part of the figure is a diagram of the spectrometer
configuration used to separate the four possible spin-dependent
scattered intensities for polarized neutrons. The Heusler(HSLR)
crystals reflect a narrow wavelength band of predominantly one
neutron spin state when magnetized perpendicular to the scattering
vector. The flat coil flippers (FLP) rotate the +(-) to the -(+)
neutron spin eigenstate, thereby allowing one or the other spin
state to be selected(43).

Table 3.3 Comparison of observed intensities (normalized to I_{-2}^{++}) flipping ratios for $[Gd(10L)/Y(10L)]_{225}$ with corresponding values calculated for the antiphase domain model described in the text. The uncertainties in the observed intensities are of the order of 5% and where indicated to be zero, to within background (43).

Reflection Order $(0002)^m$	Q (Å^{-1})	T (k)	H (Oe)	I_m^{++} OBS.	CALC.	I_m^{--} OBS.	CALC.	I_m^{+-} OBS.	CALC.	I_m^{-+} OBS.	CALC.	$R^{++/--}$ OBS.	CALC.
$\bar{4}$	1.968	75.	150.	0.046	0.008	0.050	0.011	0	0	0	0	0.93	0.73
$\bar{3}$	2.021	"	"	0	0	0	0	0.53	0.51	0.52	0.51	-	-
$\bar{2}$	2.075	"	"	1.00	1.00	1.05	1.10	0	0	0	0	0.96	0.91
$\bar{1}$	2.129	"	"	0	0	0	0	1.70	1.77	1.69	1.77	-	-
0	2.183	"	"	4.46	4.54	3.10	3.21	0	0	0	0	1.44	1.42
1	2.236	"	"	0	0	0	0	1.22	1.33	1.25	1.33	-	-
2	2.290	"	"	0.50	0.53	0.65	0.70	0	0	0	0	0.77	0.76
$\bar{4}$	1.968	79.	5.0×10^4	0.049	0.053	0.054	0.053	0	0	0	0	0.91	0.83
$\bar{3}$	2.021	"	"	0	0	0	0	0	0	0	0	-	-
$\bar{2}$	2.075	"	"	1.00	1.00	1.11	1.31	0	0	0	0	0.90	0.77
$\bar{1}$	2.129	"	"	0	0	0	0	0	0	0	0	-	-
0	2.183	"	"	5.00	4.90	1.08	0.95	0	0	0	0	4.63	5.18
1	2.236	"	"	0	0	0	0	0	0	0	0	-	-
2	2.290	"	"	0.57	0.41	0.96	0.92	0	0	0	0	0.60	0.45

to the superposition of the (002) reflection of a pure Y seed layer (amounting to about 16%) was subtracted. Further analysis of the data indicated that extinction effects were negligible. The measured intensities integrated over the entire resolution volume (with overall uncertainties of the order of 5%) were then compared to values predicted by various models for the magnetic structure. It was found that the antiphase domain model depicted in Figure 3.3.2 (in which the atomic interplanar spacing and net magnetization per plane are uniform) is a good approximation. It is assumed that the Gd moments are confined to the basal plane, even in low fields, based on the results of magnetization measurements performed on a number of samples(44). The relatively small changes in absorption and Debye–Waller factors over the limited Q–range about the (002) reflection were neglected.

Analytic expressions for the squares of the structure factors $|F_m|^2$ (which are proportional to the intensities) can be derived for this simple model:

$$|F_m^{++--}|^2 = \left| \frac{\sin m\,\pi/4}{\sin m\,\pi/40} \,(1+e^{i\pi m})\cdot \right.$$

$$\left. [b_Y + (-1)^{m/2} \,(Reb_{Gd}-iImb_{Gd} + p_{Gd}\cos\varepsilon)] \right|^2 \qquad (3.8)$$

$$|F_m^{+--+}|^2 = \left| \frac{\sin m\,\pi/4}{\sin m\,\pi/40} \,(1-e^{i\pi m}) \, p_{Gd}\sin\varepsilon \right|^2 \qquad (3.9)$$

where $b_Y = 0.775 \times 10^{-12}$ cm, $Reb_{Gd} = 0.47 \times 10^{-12}$ cm and $Imb_{Gd} = 1.2 \times 10^{-12}$ cm are the real and imaginary parts, respectively, of the nuclear scattering lengths at $k_i = 2.67$ Å^{-1} and $p_{Gd} = (0.27 \times 10^{-12}$ cm $)f\mu$, where f is the Gd magnetic form factor (assumed to be that for elemental Gd (46)) and μ is the Gd moment in number of Bohr magnetons μ_B.

However, better agreement is obtained with a more refined version of this basic model which includes both the chemical composition modulation as well as atomic–plane–spacing modulation along the c–axis. These modulations were deduced from the x–ray measurments. Furthermore a reduction in the net ordered moment of the interface planes is consistent with both magnetization (41) and magnetic x–ray scattering data (47). The intensities calculated according to this improved model are compared to the measured values in Table 3.3 at low and high magnetic fields along with the corresponding "flipping ratios" $R_m \equiv I_m^{++}/I_m^{--}$. The calculated values are those which give the best fit and are obtained for an individual interior Gd moment magnitude of 6.3 μ_B in close agreement with the measured bulk value (46), and for an angle ε (see Fig.3.3.3) of 80° at 150 Oe and zero degrees at 50 Oe. Models for other moment configurations including various helical arrangements gave significantly poorer agreement with the data.

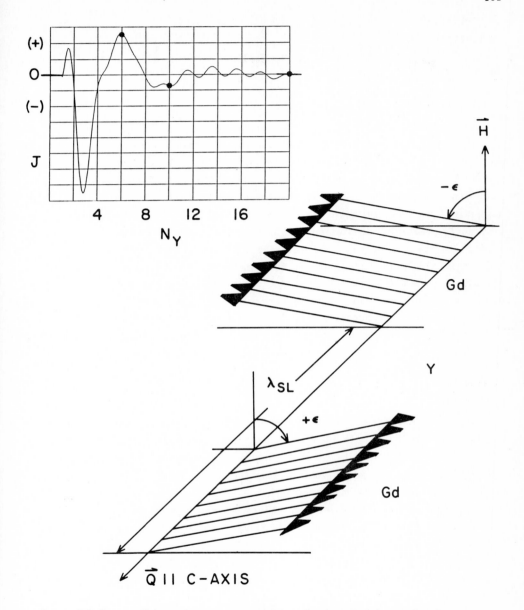

Fig. 3.3.3   Schematic representation of the antiphase domain
configuration of the Gd magnetic moments in the $[Gd(10L)/Y(10L)]_{225}$
superlattice. ε is approximately 80 degrees at 75K and 150 Oe.
The inset in the upper left-hand corner shows the results (solid
curve) of the theoretical calculation of the relative strength and
sign of the RKKY interaction as a function of N_Y that is described
in the text. The points indicate those values of N_Y for which the
magnetic structures have been determined experimentally. The
interaction is positive or ferromagnetic for $N_Y = 6L$ and negative
or antiferromagnetic for $N_Y = 10L$, in agreement with experiment.

The magnetic behavior of each of the other two samples, namely, [Gd(10L)/Y(6L)]$_{189}$ and [Gd(10L)/Y(20L)]$_{100}$, is quite striking in contrast to that of the superlattice for which N$_Y$ = 10L. Neither develops antiferromagnetic correlations down to a temperature of 10K in zero field. Both samples do, however, exhibit long-range ferromagnetic order below T$_C$.

One particularly appealing explanation for the observed oscillatory behavior is to attribute the coupling between two Gd layers across an intervening Y layer to the RKKY interaction. In order to estimate the strength and sign of this coupling, a calculation of the interaction between two Gd monolayers separated by pure Y (for which the calculated susceptibility function was assumed (48)) was performed (49). While the validity of this approximation has not been rigorously tested, it is probably the simplest physically plausible computation that can be made short of doing a superlattice band calculation. The sign and relative strength of the interaction obtained from this calculation is plotted in the inset of Fig.3.3.3 as a function of NL$_Y$. The points on the theoretical curve indicate those values of NL$_Y$ for which the magnetic structures have been determined and show that the experimental observations are indeed consistent. The calculated curve predicts that a field of the order of 6kOe would be required to bring the Gd layers of the [Gd(10L)/Y(10L)]$_{225}$ sample into parallel alignment at zero K which is in agreement with observation.

3.3.2.B Dy/Y

Dy/Y superlattices have also been studied in some detail by neutron diffraction (50-53). Bulk Dy orders antiferromagnetically at T$_N$= 178K but becomes ferromagnetic at T$_C$ = 85K. The antiferromagnetic configuration is a simple helix with propagation vector along the c* axis and with moments confined to the basal plane; the magnitude of the spiral propagation vector varies smoothly with temperature between values corresponding to a turn angle of approximately 43° at T$_N$ and 26° just above the Curie temperature (54). Dy/Y superlattices exhibit a somewhat different behavior. In superlattices with bilayers consisting of between 13 and 16 atomic basal planes of Dy and 9 to 42 planes of Y, a simple antiferromagnetic spiral in the Dy layers is observed, as in the bulk, but no ferromagnetic transition in zero field occurs down to at least 4K. What is more interesting, however, is that the helical order is coherent over many bilayer periods for samples with smaller Y layer thicknesses (e.g., 9 to 14 atomic planes of Y) as directly determined by measurements of the widths of the magnetic neutron diffraction satellites (50). This implies that the phase of the Dy spiral is propagated coherently through the intervening Y layers. For Y layer thicknesses greater than about 20 atomic planes, on the other hand, the coherence length is increasingly diminished to a single bilayer period at about a 40-plane Y layer thickness (52,53). From a quantitative analysis of the positions and relative intensities of the magnetic neutron diffraction reflections, the temperature dependence of the Dy spiral wavevector, as well as the phase change of the propagation wavevector across the Y, can be determined. It is found that the former varies smoothly, as in the bulk, but locks into a constant value at low temperatures whereas the latter is a non-integer multiple of π corresponding to a temperature-independent 51° turn angle per Y plane. It has been suggested (51,52) that the mechanism of long-range coupling is a conduction band spin density wave in both Y and Dy stabilized by the 4f spins of Dy.

3.4 CONCLUSION

To conclude, neutron diffraction is in principle a powerful technique for studying the microscopic magnetic structures of synthetic superlattices. Polarized neutrons are particularly suited for determining the detailed magnetization profile along the modulation direction in ferromagnetic multilayers and the effects of interlayer magnetic coupling. The full potential of this method should be realized as progress is made toward the growth of ideal superlattices with minimal interdiffusion and variations in the modulation period.

The authors wish to acknowledge many fruitful collaborations with our colleagues in this research. Work at Brookhaven was supported by the Division of Materials Sciences, U.S. Department of Energy, under contract No. De-ACO2-76CHO0016. Work at Tohoku University was partly supported by the Grant in Aid for Scientific Research from the Ministry of Education, Science and Culture, Japan.

REFERENCES

(1) A.J.Freeman, J.Xu and T.Jarlborg, J.Mag. & Magn. Matls. 31-34,(1983)909 and references therein.
(2) K.Binder and D.P.Landau, Phys. Rev. Lett. 52,(1984) 318 and references therein.
(3) C.F.Majkrzak, Applied Optics 23 (1984) 3524.
(4) A.M.Saxena and B.P.Shoenborn, Acta Cryst. A33 (1977) 805.
(5) E.Spiller,in AIP Conf. Proc. 75, Edited by D.T. Atwood and B.L.Henke (1981)124.
(6) T.W.Barbee,Jr.,in AIP Conf. Proc. 75, Edited by D.T.Atwood and B.L.Henke (1981) 131.
(7) B.Vidal and P.Vincent, Applied Optics 23 (1984) 1794.
(8) E.O.Wollan and C.G.Shull, Phys. Rev. 73 (1948) 830 ; C.G.Shull and E.O. Wollan, Phys. Rev. 81 (1951) 527
(9) R.Mason, E.W.J.Mitchell and J.W.White,(Eds.) Neutron Scattering in Biology, Chemistry and Physics, (The Royal Society, London) 1980.
(10) G.E.Bacon, Neutron Diffraction, 3rd Edition, Oxford University Press, London 1975.
(11) S.W.Lovesey , Theory of Neutron Scattering from Condensed Matter, Oxford University Press, London 1973.
(12) B.T.M.Willis,(Ed.) Chemical Applications of Thermal Neutron Scattering, Academic Press, New York 1979.
(13) G.Kostorz,(Ed.) Neutron Scattering, Academic Press, New York 1979.
(14) G.L.Squires, Thermal Neutron Scattering, Cambridge University Press, London 1978.
(15) H.Dachs,(Ed.) Neutron Diffraction, Springer-Verlag, Berlin, 1978.
(16) B.T.M. Willis,(Ed.) Thermal Neutron Diffraction, Oxford University Press, London 1970.
(17) I.I. Gurevich and L.V.Tarasov, Low-Energy Neutron Physics, American Elsevier Publishing, New York 1968.
(18) P.A.Egelstaff,(Ed.) Thermal Neutron Scattering, Academic Press, New York, 1965.
(19) V.F.Turchin, Slow Neutrons, Sivan Press, Jerusalem 1965.
(20) L.Koester and A.Steyerl, Neutron Physics, Springer Tracts in Modern Physics, 80, (Springer-Verlag, Berlin) 1977.
(21) N.F.Mott and H.S.W.Massey, The Theory of Atomic Collisions, 2nd Edition, Clarendon Press, Oxford, 1949.
(22) V.F.Sears, Thermal Neutron Scattering Lengths and Cross Sections for Condensed Matter Research , Atomic Energy of Canada Report AECL-8490 Chalk River, Ontario, 1984.
(23) O.Halpern and M.H.Johnson, Phys. Rev. 55, (1939) 898.
(24) R.M.Moon, T.Riste and W.C.Koehler, Phys. Rev. 181,(1969) 920.
(25) A.Steyerl, K-A. Steinhauser, S.S.Malik and N.Achiwa, J.Phys D18 (1985) 9.
(26) P.Croce and B.Pardo, Nouv. Rev. Opt. Appl. 1 (1970) 229.
(27) C.F.Majkrzak and L.Passell, Acta Cryst. A41 (1985) 41.
(28) B.Hamelin, Nucl. Instr. and Methods 135 (1976) 299.
(29) J.Schelten and K.Mika, Nucl. Instr. and Methods 160 (1979) 287.
(30) S.Yamada, T.Ebisawa, N.Achiwa, T.Akiyoshi and S.Okamoto, Annual Res. Reactor Inst. Kyoto Univ. 11 (1978) 8.
(31) C.F.Majkrzak, Physica 136B (1986) 69.
(32) R.M.Fleming, D.B.McWhan, A.C.Gossard, W.Wiegmann and R.A.Logan, J. Appl. Phys. 51 (1980) 357.
(33) A.J.Freeman, T.Jarlborg, H.Krakauer, S.Ohnishi, D.Wang, E.Wimmer and M. Weinert, Proceedings of the Yamada Conf. VII, Muon Spin Rotation, Shimoda, 1983, Hyperfine Interactions 17-19 (1984) 413.
(34) C.G.Windsor, Pulsed Neutron Scattering, Halsted Press, New York, 1981.
(35) A.Freund and J.B.Forsyth, in Ref.13.
(36) M.J.Cooper and R.Nathans, Acta Crst.A23 (1967) 357, A24 (1968) 481.
(37) Y.Endoh, J.de Physique C7 (1982) 159.
(38) Y.Endoh, N.Hosoito and T.Shinjo, J. Magn. & Magn. Matls. 35 (1983) 93

(39) C.F.Majkrzak, J.D.Axe and P.Boni, J. Appl. Phys. 57 (1985) 3657.
(40) J.Kwo, D.B.McWhan, M.Hong, E.M.Gyorgy, L.C.Feldman, and J.E.Cunningham, in Gibson and Dawson (Eds.), Layer Structure, Epitaxy and Interfaces, Conference Proceedings Vol.37, Materials Research Society, 1986, P509.
(41) J.Kwo, E.M.Gyorgy, D.B.Mcwhan, M.Hong, F.J.Di Salvo, C.Vettier and J.E. Bower, Phys. Rev. Lett. 55, (1985) 1402.
(42) S.Shinha, J.Cunningham, R.Du, M.B.Salamon and C.P.Flynn, J. Mag. & Magn. Matls. 54–57 (1986) 773.
(43) C.F.Majkrzak, J.W.Cable, J.Kwo, M.Hong, D.B.McWhan, Y.Yafet, J.W.Waszczak and C.Vettier, Phys. Rev. Lett. 56 (1986) 2700.
(44) In bulk Gd, which is a nearly isotropic ferromagnet (T_c=290K), the moments are inclined to the c-axis at an angle that is temperature dependent. For the scattering geometry with $\mathbf{Q}$ parallel to the c-axis, a component of the moment along that axis would not contribute to the scattering and cannot be conclusively ruled out without measurements of in-plane reflections (45).
(45) J.W.Cable and E.O.Wollan, Phys. Rev. 165 (1968) 733.
(46) R.M.Moon, W.C.Koehler, J.W.Cable and H.R.Child, Phys. Rev. B5 (1972) 997.
(47) C.Vettier, D.B.McWhan, E.M.Gyorgy, J.R.Kwo, B.M.Buntschuh and B.W.Batterman Phys. Rev. Lett. 56 (1986) 757.
(48) R.P.Gupta and A.J.Freeman, Phys, Rev. B13 (1976) 4376.
(49) Y.Yafet, unpublished.
(50) M.B.Salamon, S.Sinha, J.J.Rhyne, J.E.Cunningham, R.W.Erwin, J.Borchers and C.P.Flynn, Phys. Rev. Lett. 56 (1986) 259.
(51) J.J.Rhyne, R.W.Erwin, M.B.Salamon, S.Sinha, J.Borchers, J.E.Cunningham and C.P.Flynn, J. Less Common Metals, to be published.
(52) J.J.Rhyne, R.W.Erwin, J.Borchers, S.Sinha, M.B.Salamon, R.Du and C.P.Flynn, J. Appl. Phys., to be published.
(53) M.Hong, R.M.Fleming, J.Kwo, J.V.Waszczak, J.P.Mannaerts, C.F.Majkrzak, L.D.Gibbs and J.Bohr, J. Appl. Phys. to be published.
(54) M.K.Wilkinson, W.C.Koehler, E.O.Wollan and J.W.Cable, J. Appl. Phys. 32, (1961) 48S.

Chapter 4

Mössbauer Spectroscopic Studies on Superlattices

T. SHINJO
Kyoto University

4.1 INTRODUCTION

In 1958, Mössbauer reported that nuclei in solids may emit and absorb gamma-rays without any recoil energy being transferred to the lattice(1). Since the energy width of the gamma-ray is very narrow, typically on the order of 10^{-9} eV, hyperfine interactions of the nucleus can be measured through the "Mössbauer effect". Although the experiment was initiated in the field of nuclear physics, Mössbauer spectroscopy as an experimental tool was soon introduced in solid state physics and then in various fields such as chemistry, biology and metallurgy. Hyperfine interactions of the nucleus provide us with direct, microscopic information on the electronic structure of the relevant atom. As in the case of NMR, Mössbauer spectroscopy is representative of experimental techniques which utilize nuclei as microprobes in condensed matter.

The Mössbauer effect can occur in more than 100 nuclear transitions of more than 40 elements. The number of technically easy cases, however, is fairly limited. Fortunately, ^{57}Fe is the most suitable nuclear species for the Mössbauer effect measurements. As is well known, Fe is very often the leading actor in the field of magnetism and also plays important roles in chemistry. Since Fe is one of the most abundant metals on earth, its importance in geology and mineralogy is evident. The central atom of hemoglobin in the human body is also Fe. By using Mössbauer spectroscopy, a great number of investigations have been carried out(2). About two-thirds of the publications, however, are concerned with ^{57}Fe, and a half of the other third, with ^{119}Sn. This fact proves how useful and, at the same time, how easy it is to take ^{57}Fe Mössbauer measurements.   At least as far as ^{57}Fe and ^{119}Sn are concerned, the experimental techniques have been well established. Since an experimental setup and a standard gamma-ray source are commercially available, one can obtain spectra without any difficulty. In addition, the spectra for these two nuclear species are rather simple and the meaning of the result can often be understood intuitively.

Before the usefulness of Mössbauer spectroscopy for the study of superlattices is discussed, a general explanation of the information derived from ^{57}Fe Mössbauer spectra is presented for readers who are not familiar with this method. Because of limited space, the physical principles of the Mössbauer effect and instrumental descriptions are omitted here. For details of the physical background of the Mössbauer effect, one may refer to textbooks(3).

Section 4.3 of this chapter summarizes Mössbauer spectroscopic studies on the magnetic properties of ferromagnetic metal interfaces. The need for superlattice samples has been realized in the course of the extensive study of interface magnetism. A knowledge of interfaces, on the other hand, is useful for studies on microscopic structure of superlattices. Applications of Mössbauer spectroscopy to the study of superlattices are described in the last section.

4.2 INFORMATION FROM ^{57}Fe MÖSSBAUER SPECTRA

For the Mössbauer effect measurements of the 14.4keV gamma-ray from ^{57}Fe nucleus, the radioactive isotope ^{57}Co is used as the parent nucleus. Having a lifetime of 270 days, ^{57}Co decays first to the second excited state of ^{57}Fe, then immediately to the first excited state, and finally to the ground state while emitting the 14.4keV gamma-ray. Usually a standard ^{57}Co source is used which is embedded in a cubic, non-magnetic metal matrix such as Rh or Cu with an initial activity of 10 ∿ 100mCi. The situation of the ^{57}Fe nucleus in an absorber is the object of the Mössbauer spectroscopic study. ^{57}Fe is a stable isotope found in natural Fe in an abundance of 2.2%.

The energy schemes and the absorption spectra of typical cases are shown in Fig.4.2.1. For special purposes, samples including ^{57}Co are prepared and the situation of the source nucleus is investigated by using a standard absorber such as a ^{57}Fe-enriched stainless steel foil. The results from these source experiments can be interpreted analogously with the absorber measurements. If the energy of the gamma-ray from the source is exactly the same as the energy between the ground state and the excited one in the absorber, the Mössbauer transition (recoilless gamma-ray resonant absorption) occurs and then the counting rate of transmitted gamma-ray should decrease. In a normal setup, the standard source is vibrated at a frequency of several Hz, in order to give a slight energy shift to the gamma-ray due to the Doppler effect. Gamma-rays transmitted through the absorber are counted by a proportional counter and then stored in a multichannel analyzer. Since the sweeping of the channels repeats

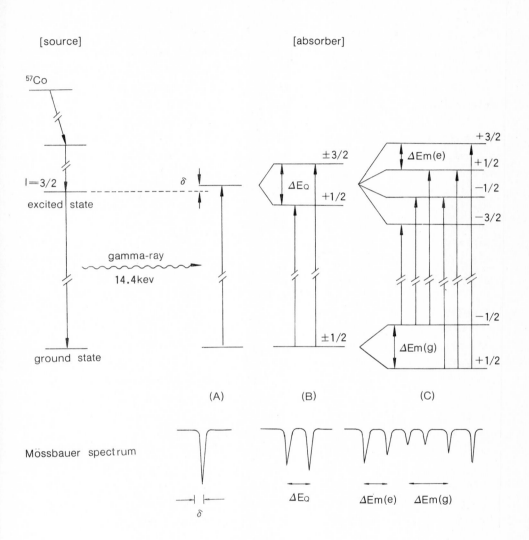

[source]　　　　　　　　　　[absorber]

^{57}Co

$I=3/2$
excited state

gamma-ray
14.4kev

ground state

δ

ΔE_Q
±3/2
+1/2

±1/2

+3/2
+1/2
$\Delta Em(e)$
−1/2
−3/2

−1/2
$\Delta Em(g)$
+1/2

(A)　　　　　(B)　　　　　(C)

Mössbauer spectrum

δ　　　　ΔE_Q　　　$\Delta Em(e)$　　$\Delta Em(g)$

Fig. 4.2.1 Schematic diagram of the nuclear energy shift and splitting
of ^{57}Fe.
(A) EFG=0, H_n=0, (B) EFG≠0, H_n=0, (C) EFG=0, H_n≠0.
δ: isomer shift,
ΔE_Q: quadrupole splitting,
$\Delta E_m(g)$: Zeeman splitting of the ground state,
$\Delta E_m(e)$: Zeeman splitting of the excited state.

synchronously with the source vibration, each channel in the analyzer corresponds to a certain Doppler velocity. If the Mössbauer effect occurs, the counts of certain channels should be less than those of the background level. If the source vibration is parabolic in mode, acceleration is constant and velocity changes linearly against time. The vertical axis of Mössbauer spectra refers to the counting rate of gamma-rays, while the horizontal one refers to the energy shift in the gamma-ray due to the Doppler effect. For convenience sake, this shift is normally expressed by the unit of velocity, $mm \cdot sec^{-1}$, instead of an energy unit such as eV.

If there is neither an electric field gradient(EFG) nor a magnetic hyperfine field at the ^{57}Fe site, the spectrum should be a single line as shown in Fig.4.2.1(Case A). If the surrounding charge distribution is not in cubic symmetry, an EFG occurs and the spectrum shows a doublet due to a quadrupole interaction (Case B). A symmetric six-line pattern is observed when a magnetic hyperfine field is present but without an EFG (Case C). Mössbauer spectra mainly provide information on (i) the isomer shift, (ii) magnetic hyperfine field and (iii) quadrupole interaction. Below, each Mössbauer parameter is explained briefly. Although the recoilless fraction has useful information for the study of lattice dynamics, a description is not included here.

(i) Isomer shift

Isomer shift is defined as the center of gravity of the spectrum. In the case of a single line spectrum, the isomer shift can easily be estimated from the position of the absorption peak. This shift arises from an interaction between the nucleus and the electric field at the nuclear site, which is the electronic charge density, $|\psi(0)|^2$. As the size of the nucleus (positive charge) changes during the nuclear transition, the electric field produces an energy shift at the nuclear level. Usually the effect of electrons on the nuclear energy level is extremely small, 10^{-12} for instance, and cannot be distinguished by usual experimental techniques though it can be observed through the Mössbauer effect.

If the Mössbauer atoms in the source and the absorber have different chemical surroundings, the Mössbauer line is shifted by the amount, δ,

$$\delta = -\frac{2\pi}{5} Ze^2 \ \{|\psi(0)|^2_a - |\psi(0)|^2_s\}\{<R^2_e> - <R^2_g>\} \tag{4.1}$$

where "a" and "s" stand for absorber and source, respectively. The atomic number of the Mössbauer atom is Z and $<R^2_e>$ and $<R^2_g>$ are the mean square radii of the nucleus in the excited and ground states, respectively. The second term is the difference in nuclear size and is constant for each nuclear transition. Thus, the value of the isomer shift is proportional to the difference in the electronic charge densities between the source and absorber nuclei. Although the charge densities measured from the isomer shift are not those of valence electrons but of core electrons, the values reflect the situation of valence electrons to a reasonable extent. Figure 4.2.2 shows the relation between the isomer shift and the valence state of Fe. Each ionic valence state has a certain range. Therefore, the isomer shift is very useful for identifying the valence state. At a very early stage, attempts were made to analyze the electronic configuration from the isomer shift value and many studies have followed(4). However, generally speaking, quantitative analyses of isomer shift values are not simple. Besides the isomer shift due to the electronic state, a temperature dependent shift arises from the second order Doppler effect due to lattice vibration. This shift is usually minor but should be corrected for a quantitative analysis.

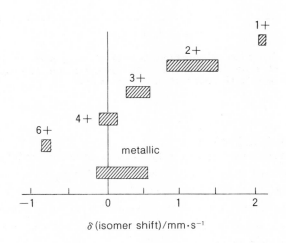

Fig. 4.2.2   Isomer shift of ^{57}Fe in various valence states (high spin states) relative to the center of an α-Fe metal spectrum at room temperature.

In order to avoid confusion from the differing source matrices, isomer shift values are expressed relative to a standard absorber. In most cases, the center of the spectrum from a pure α-Fe foil at room temperature is used as the zero velocity point. The peak positions of the pure Fe spectrum are precisely known (Fig.4.2.3). By referring to this spectrum, the relation between channel and velocity can be calibrated and the linearity of the velocity scale checked.

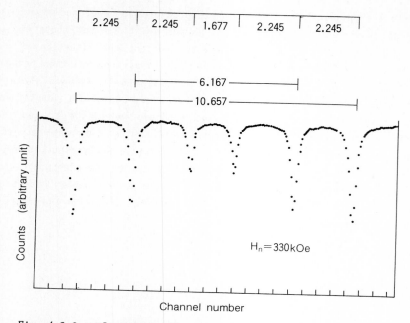

Fig. 4.2.3   Mössbauer absorption   spectrum of   a pure   α-Fe
metal foil at room temperature.   The numbers for the peak
separations are in the unit of velocity, mm·sec.$^{-1}$

(ii) Magnetic hyperfine interaction

The hyperfine field has been widely recognized as providing very useful
information for microscopic studies in the field of magnetism.   If a nucleus
experiences a magnetic field, the interaction between the field and nuclear
magnetic dipole moment causes a Zeeman splitting of each nuclear level of spin I
into 2I+1 sublevels.   In the case of ^{57}Fe, six transitions are allowed and
therefore six peaks appear in the spectrum (Case C in Fig.4.2.1).   The magnitude
of splitting in the Mössbauer pattern is proportional to the hyperfine field.
If the nuclear magnetic moment of the ground state is known from the NMR result,
the hyperfine field can be estimated accurately.   In a magnetic atom, an
unpaired electron spin produces a very large magnetic field at the nuclear site,
which is, for instance, 330kOe in ferromagnetic Fe metal at room temperature.
This results from the spin polarization of core electrons.   Due to an exchange
interaction with d-electrons, the radial distributions of spin-up and spin-down
electrons in core s-orbitals are slightly different.   Although the degree of
polarization is rather small, the induced magnetic field is quite large since
the core s-electrons have finite densities at the nucleus (Fermi contact
interaction).

The relation between the local magnetic moment and hyperfine field is rather complicated and, in general, quantitative analyses are difficult. Theoretical studies suggest that the dependence of the hyperfine field on the electronic structure is delicate(5). Nevertheless, magnetic hyperfine field measurements have contributed greatly to the understanding of magnetic properties of materials. A rough proportionality between the hyperfine field and local magnetic moment has been found in many cases of the Fe atom in ferromagnetic alloy systems. A phenomenological analysis can be made by assuming that the hyperfine field is the sum of the contributions from the atom itself and from the environment, an example of which is given in Chapter 5. In magnetically ordered systems, the temperature dependence of the hyperfine field is proportional to that of the local magnetization. The magnetic transition temperature is therefore determined from the collapsing of the magnetic hyperfine structure in the spectrum. The value of the hyperfine field which corresponds to the full magnetic moment is estimated by extrapolating it to zero temperature. In ionic compounds, Fe^{3+} ions show similar hyperfine fields ranging from 500 to 600kOe. On the other hand, hyperfine fields of Fe^{2+} ions differ widely since the orbital angular momentum contributes significantly.

The presence of a magnetic hyperfine structure in a Mössbauer spectrum is undoubted evidence of the existence of a magnetic order. Hence, Mössbauer spectroscopic measurements are especially useful for studying antiferromagnetic materials. However, strictly speaking, the observed hyperfine structure means that the magnetic field is spatially stable during the nuclear Larmor precession time. (Usually the Larmor precession time is shorter than the intrinsic observation time of the Mössbauer effect, which is the mean lifetime of the excited state, 10^{-7}sec for ^{57}Fe). Accordingly, if the electronic spin system takes part in a dynamical phenomenon with the relaxation time of this region, the result of the Mössbauer measurements should be carefully interpreted. Superparamagnetism, for instance, may occur in fine particles or extremely thin films. Because the number of magnetic spins in the system is not sufficiently large, the bulk magnetization as a whole begins to fluctuate thermally even though the spins are coupled with each other. If the relaxation time of superparamagnetism becomes shorter than the nuclear Larmor precession time, the magnetic hyperfine structure in the spectrum begins to collapse at a temperature much lower than the Curie temperature(6).

On the contrary, a paramagnetic spin system may exhibit a hyperfine structure if the relaxation time is long enough. Usually the relaxation time of the paramagnetic electron spin is much shorter than the Mössbauer observation time and hence no hyperfine structure is observed though each atom has a local magnetic moment. The situation to observe a paramagnetic hyperfine structure is realized in insulating systems only if the concentration of magnetic spins is

114

very low. An example is Al_2O_3 including $1\%Fe_2O_3(7)$. Thus it is not difficult
to distinguish the hyperfine structure of a paramagnetic state from that of a
magnetically ordered state.

The direction of magnetization may be speculated from the intensity ratio
of the six lines in a magnetically split Mössbauer pattern. In the case of
ideally thin absorbers, the relation between the intensity ratio and the angle
between the gamma-ray beam and direction of the magnetic field, θ, is as
follows: If the intensity of six lines is denoted as 3:X:1:1:X:3, the
dependence of X on θ is expressed as $4\sin^2\theta/(1+\cos^2\theta)$. For a powder sample
where θ is randomly distributed, the ratio becomes 3:2:1:1:2:3. If the
magnetization is perfectly oriented in the direction of the gamma-ray, or if a
strong magnetic field is applied in this direction, an intensity ratio of
3:0:1:1:0:3 would be obtained. If X is non-zero even when the applied external
field is strong enough for the saturation of the bulk magnetization, the spin
structure is suggested to be non-collinear. It should be noted that the sign of
the hyperfine field is usually negative, which means that the direction of the
magnetic field at the nucleus is opposite to that of the electron spin.
Therefore, the observed hyperfine field is reduced when an external field is
applied in ferromagnetic cases.

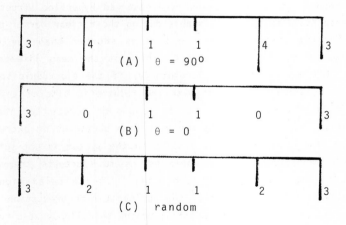

Fig. 4.2.4 Effect of the orientation of magnetic field with
respect to the gamma-ray vector on the relative intensities
of a ^{57}Fe six-line pattern.
(A) θ = 90°,
(B) θ = 0°,
(C) powder pattern.

Compared to NMR, as an experimental tool to measure hyperfine fields, the resolution of Mössbauer spectroscopy is worse and selection of the nuclear species is limited. A merit of Mössbauer spectroscopy, however, is the ease of taking measurements in the cases of ^{57}Fe and ^{119}Sn. Furthermore, measurements can be taken in a wider temperature range. Mössbauer spectroscopy is especially useful for examining a sample with multiphases having different magnetic characters (a mixture of ferromagnetic and para or antiferromagnetic phases). Therefore, NMR and Mössbauer techniques should be used complementarily.

(iii) Quadrupole interaction

If the surrounding charges are not in cubic symmetry, the interaction of the nuclear quadrupole moment with the EFG may also split the energy levels. The nuclear quadrupole moment reflects the deviation of the nuclear charge distribution from spherical symmetry. Nuclei with I=0 or 1/2 are spherically symmetric and have no quadrupole moment; hence the ground state of ^{57}Fe with I=1/2 cannot exhibit a quadrupole splitting. On the other hand, the excited state of ^{57}Fe is split into levels, $\pm 1/2$ and $\pm 3/2$ (Case B in Fig.4.2.1). The EFG originates from either electrons directly associated with the ^{57}Fe atom or charges on surrounding atoms. If the distribution of valence electrons is not spherical, a large EFG may result. An Fe^{3+} ion has five 3d electrons which spherically occupy all 3d orbitals, and hence quadrupole splittings observed for Fe^{3+} compounds are not very large, ranging from 0 to $1.0mm\cdot s^{-1}$. In contrast, an Fe^{2+} ion has one more valence electron which produces a significantly large EFG and therefore the observed quadrupole splittings for Fe^{2+} compounds typically reach $3mm\cdot s^{-1}$. In metallic systems, the electronic structures of individual atoms are regarded as nearly spherical and the quadrupole splittings are usually less than $0.5mm\cdot s^{-1}$. Quadrupole splittings thus indicate the degree of deviation of the local charge distribution from cubic symmetry and are useful for identifying crystallographic sites.

The intensity ratio of two Mössbauer lines split by a quadrupole interaction depends on the direction of the gamma-ray beam. In the case of uniaxial EFG, the ratio is expressed as $(3+3\cos^2\theta):(2+3\sin^2\theta)$, where θ is the angle between the principal axis of EFG and the direction of the gamma-ray propagation. If θ is 0 or 90°, the ratio becomes 3:1 or 3:5. A randomly oriented sample should show a 1:1 ratio, regardless of the nature of EFG, though actual Mössbauer spectra for powder samples sometimes show slight deviations from the exact 1:1 ratio. An asymmetric doublet may be induced by the anisotropic lattice vibration (Karyagin-Goldanskii effect)(8) or by the magnetic relaxation with a specific relaxation time(9). However, in almost all cases, the reason for the asymmetric doublet is a partial preferred orientation of particles (or grains) in the sample.

In a magnetically ordered material, a magnetic hyperfine interaction coexists with a quadrupole interaction and then the situation is not always simple. Unless the principal axis of EFG coincides with the direction of the magnetic field, the energy levels of the excited state become mixed and in principle eight transitions become possible, instead of six in a purely magnetic case. However, a magnetic hyperfine interaction is usually dominant and therefore a quadrupole interaction appears only as a small perturbation. In other words, the six Mössbauer lines split by the magnetic hyperfine interaction are systematically shifted by the quadrupole interaction. The shifts are determined by the magnitude of the quadrupole interaction and the relative angle between the magnetic field and EFG principal axis. If the EFG is well defined, the magnetic structure may be speculated from the combined effects of the magnetic and quadrupole interactions. A successful example is $\alpha-Fe_2O_3$, whose spin flip transition at about 270K has been unambiguously identified from powder Mössbauer patterns(10).

4.3 INTERFACE MAGNETISM

(i) Problems of surface and interface magnetism

Magnetic behaviors near a surface or interface may differ in many respects from those of the interior(11). First of all, local magnetic moments at surface sites at zero temperature should be considered. In metallic systems, the magnetic moment is sensitive to the local environment. The discussion on surface magnetic moment evolved from a proposal of the magnetically dead layer model by Liebermann et al.(12). This group measured the magnetization of electrodeposited thin films of Fe, Co or Ni as a function of thickness and explained the results by assuming that the first surface layer of such ferromagnetic metals has no magnetic moment. The situation of an electrolytically prepared surface is far from that of a vacuum surface and is hard to characterize. There are indeed many ambiguities in drawing such a conclusion. As a matter of fact, almost all experimental results on vacuum surface and also on various kinds of interface are against the dead layer model. It is therefore believed that magnetically dead layers do not exist in usual surfaces or interfaces of ferromagnetic 3d metals or alloys. The concept of the dead layer is, however, well defined and the use of this term has been popular in the course of discussion.

Theoretical studies also do not support the existence of the dead layer in vacuum surfaces, but rather suggest some increase of the local magnetic moment. Extensive theoretical calculations of the electronic and magnetic structures of ferromagnetic metal surfaces have been carried out by Freeman's group(13). The details of their work relevant to superlattices are introduced in Chapter 8. According to Ohnishi et al.(14), there seems to be an increase of the magnetic moment at the vacuum surface of (100) plane in ferromagnetic bcc Fe metal. A qualitatively similar result was also reported by Victoria et al.(15). Theoretical studies suggest that the magnetic surface effect, an increase of spin polarization in the above case, is only noticeable at the top surface layer and the anomaly diminishes within a few atom layers.

In conventional experiments to study the magnetic surface effects where very thin films have been used as the samples, information on the surface is derived from the thickness dependence of some macroscopic physical quantity. For example, Liebermann et al. measured the bulk magnetization as a function of thickness(12). However, if the film is thin enough to reveal a surface effect, the thickness effect also becomes significant. In such experiments, moreover, the results depend crucially on the accuracy of thickness estimation. Therefore, it is desirable to use a three-dimensional crystal or film thick enough to avoid any thickness effect for the study of surface or interface. Then, an experimental tool for a surface magnetism study should have sufficient sensitivity to recognize a fractional change in one monolayer at the surface of a three-dimensional sample. The applicability of Mössbauer spectroscopy to such problems is introduced later and compared with other techniques.

If the ground state surface magnetic structure has been understood, the next problem to be studied is the temperature dependence of the local magnetization at the surface layer. In all kinds of magnetic materials, the temperature dependence of the magnetization at the boundary between non-magnetic substances may be different from the interior behavior. At a surface site, the molecular field is supposed to be smaller than the bulk value since the number of magnetic neighbors is less. The exchange interaction in the top surface layer, or between the top and second layers, may also be somewhat different. Even when the surface magnetic moment does not change, the temperature dependence of surface magnetization can be different from the bulk curve. In ionic systems, the surface magnetic moment itself probably does not differ greatly from the normal value, but an effect of the surface may be revealed in the temperature dependence of surface magnetization.

From the early days many calculations for thin films and also for surfaces have been carried out under the assumption that all local magnetic moments are the same, well-defined values, even those at the top surface layers. Binder and his co-workers practiced Monte Carlo calculations for Ising or Heisenberg spin systems and studied the magnetic behaviors near the surface region(16). In general, at a lower temperature region, surface magnetization drops faster than the bulk curve with the increase of temperature. The surface effect, namely, the deviation from the bulk curve, penetrates into deeper regions with the increase of temperature. Below Curie temperature, in a rather wide region, the temperature dependence exhibits a linear-like behavior. In the calculations, the exchange interaction is a changeable parameter and a drastic surface anomaly is predicted if the surface exchange interaction is assumed to be greatly different from the interior value. However, a too large difference is not realistic.

Recently Hasegawa did a theoretical study of the magnetic properties of transition metal surfaces at finite temperatures based on the itinerant electron model(17). The result on Fe(100) surface is similar to an experimental one which is described later. An interesting result has been obtained for Cr(100); the surface has a rather large magnetic moment and the surface magnetic order persists at a temperature much higher than the bulk Néel Temperature(18).

Not only the magnitude of spin polarization at a surface site atom but also the preferred direction of the surface magnetic moment is interesting. The discussion on surface magnetic anisotropy was initiated many years ago by Néel(19). At a surface site, the local symmetry is lower than that in the bulk and therefore anisotropic interaction, which may be canceled by the symmetry in the interior, does not disappear at the surface. The quantization axis is in a different direction at a surface site, probably perpendicular to the plane, and spin orbit interaction may become significant. At least qualitatively it is not difficult to imagine the existence of surface anisotropy so that the spontaneous magnetization is oriented along the surface normal. Such a mechanism exists only at the top surface layer. Therefore, it may be appreciable in an extremely thin film but is negligible in a three-dimensional crystal. According to Gradmann and his co-workers, the spontaneous magnetization is perpendicular to the film plane in the case of ferromagnetic FeNi alloy films and also in Co metal films only if the thickness is less than a few atomic layers(20). As is described shortly, the Mössbauer results reveal that the spontaneous magnetization in an Fe monolayer sandwiched in-between Mg layers is also along the film normal. These results are consistent with a recent theoretical study which suggests that the perpendicular magnetization is stable in ferromagnetic metallic monolayers(21).

Surface anisotropy for ionic magnetic materials has also been argued. For instance, Morrish and Haneda have measured the Mössbauer spectra for ferrimagnetic oxide particles by applying a magnetic field strong enough for saturation and found that the no.2 and no.5 lines remain with certain intensities(22). They claimed that this is evidence of spin canting at the surface region due to the surface anisotropy.

Strictly speaking, surfaces in real samples are not perfect. Even the vacuum surface of a single crystal is not completely atomically flat and includes some chemical impurities. The situations of interfaces are more complicated, because of interdiffusion, compound formation, strain, and various lattice defects. Since the magnetic properties of the interface depend delicately on the situation, the preparation of high quality samples is essential for interface magnetism study, especially when the experimental results are to be compared with theoretical predictions. Conversely, the structure or chemical profile of the interface is reflected in the magnetic properties rather sensitively. Interface magnetic properties provide unique information on the interface characterization. Some examples are presented in this chapter and the next.

(ii) Application of Mössbauer spectroscopy

The only conditions required of Mössbauer samples are that they be solid and contain a certain amount of the Mössbauer isotope. Hence Mössbauer spectroscopy has been widely utilized to study fine particles, thin films and amorphous materials. If the particle size or film thickness is very small, the surface or interface effects should become appreciable in the spectra.
Indeed it is common to use fine particles or thin films as the samples for the investigation of surface phenomena. However, it is more desirable to use a sample having a three-dimensional size to examine the surface or interface, because then the size effects are eliminated and the intrinsic surface or interface effects can be observed.

Mössbauer spectroscopy can be used specifically for surface and interface studies in three ways:
(a) conversion electron spectroscopy
(b) gamma-ray source spectroscopy
(c) absorption spectroscopy using surface-selectively enriched samples.
Studies of (a) and (b) are briefly described below before proceeding to a discussion of (c), which is the main topic of this subsection.

Most Mössbauer experiments adopt a transmission mode in which the gamma-rays from a source are counted after passing through an absorber, but a scattering mode is also possible. If the scattered secondary radiations are counted, an increase in counts is observed at the Doppler velocity where the Mössbauer effect occurs(Fig.4.3.1). Among the secondary emissions, conversion electrons are much more abundant than gamma-rays. Mössbauer measurements by collecting the scattered conversion electrons are called conversion electron Mössbauer spectroscopy (CEMS). Since the penetration length of the electrons is limited, the depth range that can be observed by this technique is about 1,000Å. This depth range is too large to catch surface phenomenon at just a few surface atom layers. The usefulness of this technique has been recognized in the areas of catalysis and corrosion. Because gas-flow counters are used in ordinary CEMS measurements, clean surface experiments are impossible and the sample temperature cannot be easily changed.

A more advanced method is the depth-selective CEMS measurement (DCEMS). Electrons traveling through condensed matter lose energy according to the length of travel. If the energy of an emitted electron is resolved by using an appropriate electron spectrometer, we can judge the depth from which the electron emerges by assuming that the energy loss is proportional to the escape depth. For instance, information from near the top surface layer may be obtained by collecting electrons without any energy loss. Surface studies by means of DCEMS have been attempted by several groups. The measurement of (100) surface of a single crystal Fe in UHV at low temperature was very recently reported by Kiauka et al.(23). In order to enhance the surface selectivity, an angle resolution technique was also adopted. A Monte Carlo simulation was utilized to explain the relation between the electron energy loss and escape depth. They claimed that the depth resolution is sufficient to distinguish a hyperfine field at the top surface layer, if the value is anomalous. However, anomalous hyperfine fields were not detected in the spectra and they concluded that the surface magnetic moment is not significantly different from the standard value. By this DCEMS technique, magnetic behaviors of vacuum surfaces can be studied, even though there are many technical difficulties in getting enough resolution to investigate the behavior of the top surface layer. In principle, an external magnetic field cannot be applied, which is a demerit of this method.

As in the case of absorbers, Mössbauer gamma-ray sources can also be the samples to be investigated. For source spectroscopy, a standard absorber with no hyperfine splitting is utilized (stainless steel foil and $K_4Fe(CN)_6$ powder are conventionally used). Although the preparation of the source sample which involves handling of a radioactive isotope is somewhat troublesome, there are several merits to source spectroscopy. The minimum activity of ^{57}Co required for the Mössbauer source is on the order of 10 μCi, which is much less than the amount of one monolayer in an area of $1cm^2$. If ^{57}Co atoms are uniformly deposited on a crystal surface, the Mössbauer emission spectra will give us information on the top surface layer of the bulk crystal. More than ten years ago, the author's group reported some results on the surfaces of Fe and Co where ^{57}Co atoms were deposited by electrolytic means(24). The spectra at low temperatures show only ferromagnetic absorptions, indicating that the magnetically dead layer does not exist. From the Mössbauer spectra it is clear that the ^{57}Fe atoms transformed from ^{57}Co are in metallic states. However, the microscopic structure of an electrolytically prepared surface is very complicated and the local atomic configuration cannot be determined. At least it is worthwhile to note here that the Mössbauer results on "electrodeposited" surfaces are against the existence of dead layers, because the dead layer model

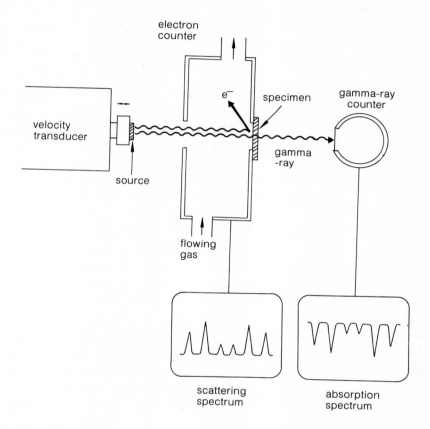

Fig. 4.3.1 Geometry for simultaneous Mössbauer measurements
in two modes, scattering and transmission.

was originally proposed from the experimental results on "electrodeposited" thin films by Liebermann et al.(12).

In order to obtain a better-defined surface, vacuum deposition of ^{57}Co is preferable to electrodeposition. A few attempts have been made to study the behaviors of ^{57}Co impurities in non-magnetic metal surfaces(25). However, successful measurements on magnetic surfaces have not yet been reported. While this technique is unique to the study of the surface state of a monolayer on a bulk crystal, the experimental difficulties seem to be greater than originally supposed. As a matter of fact, there has been no publication in this field in the past decade.

A very efficient method for studying the interface is to use surface-selectively enriched samples as the Mössbauer absorbers. Since the natural abundance of the Mössbauer isotope, ^{57}Fe, is about 2%, Mössbauer spectroscopy can be applied to samples containing not less than 10% natural Fe. By synthesizing samples from pure ^{57}Fe, the minimum content of Fe may be reduced to one-fiftieth. Enriched ^{57}Fe and ^{56}Fe isotopes are commercially available, though they are very expensive. If ^{57}Fe atoms are located only on the surface, the Mössbauer spectra will reveal the situation of the surface of a three-dimensional sample. This idea was first proposed by van der Kraan, who has attempted to coat the surface of α-Fe$_2$O$_3$ particles with a ^{57}Fe layer(26). Succeeding him, Haneda and Morrish have also tried the ^{57}Fe-coating on γ-Fe$_2$O$_3$ particles(21). A metallic interface study was initiated by Lauer et al. with successive depositions of natural Fe, ^{57}Fe and Cu in UHV(27).

If the sample base is prepared with ^{56}Fe, instead of natural Fe, the surface-selectivity should be enhanced. Figure 4.3.2 shows typical structures of samples prepared for an interface study of metallic Fe. By UHV deposition or some other technique, sufficiently thick ^{56}Fe, very thin ^{57}Fe and a non-magnetic material are deposited in turn. To study a magnetic/magnetic interface, another magnetic substance should be deposited instead of a non-magnetic one. Normally an Fe film thicker than 50Å is a continuous layer and exhibits the bulk hyperfine field. If the structure of Case A in the figure is realized as designed, the Mössbauer probe, ^{57}Fe, is located only in the topmost surface region of the Fe metal 103Å thick. For a depth-profiling measurement, another ^{56}Fe layer of a certain thickness may be deposited, such as in Case B, and thus the depth range of the ^{57}Fe layer can be varied as desired. If a sample without coating is exposed to air (Case C), the initial stage of surface oxidation can be studied.

Mössbauer transmission spectroscopic measurements require a certain Fe thickness in the absorber sample, such as 5 μm for natural Fe or 1,000Å for pure ^{57}Fe. When the thickness of a single ^{57}Fe layer in the sample is only 10Å, the sample pieces should be stacked up more than 50 times. For thinner layers, the deposition process shown in Fig.4.3.2 (Case A or B) should be repeated several times before stacking. As for the stacking of the films, single crystal substrates are not convenient, but polymer films such as mylar or kapton are suitable substrates, since they are easy to cut or fold and are also very transparent to the gamma-ray. Structural characterization such as shown in the figure is generally difficult, even when the samples are prepared on single crystal substrates. However, from our experience with multilayered samples, we believe that the structures can be synthesized almost exactly as designed. As described in Chapter 2, UHV deposited Fe layers usually have bcc structures unless they are extremely thin. Even when the substrate is amorphous, the Fe layer has a [110] texture structure. The size of each grain, however, is fairly small. The ^{57}Fe deposition on ^{56}Fe surface no doubt seems to be uniform. The diffusion of ^{57}Fe cannot be inhibited completely but is not serious as far as the substrate temperature is low. The chemical situation of the interface can be checked from the Mössbauer results, that is, whether interdiffusion or compound formation between Fe and the coating material has taken place. Also whether or not the surface has been contaminated, such as by oxidation, can be determined.

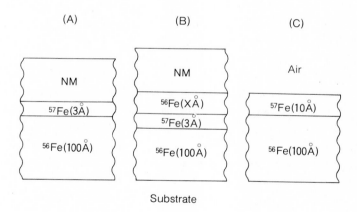

Fig. 4.3.2   Cross sections of surface-selectively enriched
samples. (NM, a non-magnetic material)

Figure 4.3.3 shows the results of the interface between Fe and V, which is an example of a magnetic/non-magnetic interface study(28). The [57]Fe-probing layers are located in three different depth ranges, 0-3.5, 3.5-7 and 7-10.5Å, from the interface boundary. The Mössbauer spectrum at 4.2K for the depth range of 7-10.5Å is a normal six-line pattern very similar to that of pure α-Fe. Regarding magnetism, interface effects do not extend to this depth. The spectrum for 3.5-7Å is also a six-line pattern but the line widths are somewhat broader, suggesting the existence of minor components with reduced hyperfine fields. The intensity ratio of both spectra is 3:4:1:1:4:3, and therefore the direction of the spontaneous magnetization is proved to be in-plane. A remarkably different spectrum is obtained from the topmost region, 0-3.5Å, where the interface effect is obviously enhanced. The broad spectrum suggests a wide distribution of the hyperfine field and the estimated distribution curve is also shown by the solid line. Here, for the sake of simplicity, a constant isomer shift and no quadrupole interaction are assumed and the ratio of the six lines is 3:4:1:1:4:3. Although the distribution is rather wide and continuous, two peaks are distinct, approximately at 240kOe and 320kOe. Since the thickness of the [57]Fe layer, 3.5Å, roughly corresponds to two atom layers, the two hyperfine fields may be attributed to the two interface layers; 240kOe corresponds to the top atom layer and 320kOe, the second one. The hyperfine field of the second layer is already very close to the bulk value, 340kOe, but the top layer has a significantly smaller field. The results can be summarized as follows: at the surface of Fe metal facing V, all Fe atoms are ferromagnetic and an interface magnetic effect is only appreciable at the top layer, where the magnetic moment is partially reduced. The average decrease of the magnetic moment due to the interface effect is about 30%, if the magnetic moment is proportional to the hyperfine field. Although this assumption is very crude, it should be mentioned that the results from neutron diffraction on Fe/V superlattices are accounted for by this model(28).

Although this model can give us an idea of an interface as a zeroth order approximation, it is apparently too naive. For a more realistic model, surface roughness and intermixing should be taken into consideration. Otherwise the observed wide distribution of the hyperfine field cannot be explained. An analysis which assumes intermixing has occurred was carried out to interpret the [51]V NMR results on Fe/V superlattices, the details of which are presented in the next chapter. The observed distribution of the V hyperfine field is satisfactorily reproduced by assuming an intermixing ranging over three atom layers. On the other hand, an analysis of Fe hyperfine field distribution in Fe/V superlattices which also takes intermixing into consideration was attempted by Jaggi et al., who used both the data of their CEMS measurements(29) and those of Ref.28. The chemical profile of the Fe/V interface thus estimated agrees

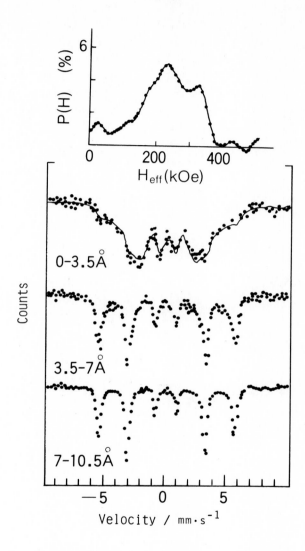

Fig. 4.3.3 Mössbauer absorption spectra at 4.2K of samples to study the Fe interface in contact with V. The depth ranges observed from the interface boundary are 0-3.5Å, 3.5-7Å and 7-10.5Å, respectively. The solid line in the figure is the curve calculated with the distribution of the hyperfine field shown separately in the above portion of the figure(28).

very well with that from the results on V. Although these two analyses were
carried out independently using different microprobes, ^{51}V and ^{57}Fe, the
obtained results are in good agreement, and therefore the conclusion that the
range of intermixing is within three atom layers can be regarded as reliable,
even quantitatively. It is a merit of Fe/V superlattice samples that the two
nuclear species, Fe and V, can be utilized complementarily as microscopic
probes. Hyperfine field distribution is significantly influenced by intermixing
in a limited range and also by surface roughness. In other words, hyperfine
field measurements are very sensitive tools for the study of interface
structures.

The situation of the Fe/Mg interface is somewhat different(30). In
contrast to Fe/V, which can form solid solutions, Fe and Mg are insoluble even
in the liquid state. The Mössbauer spectra shown in Fig.4.3.4 are for a

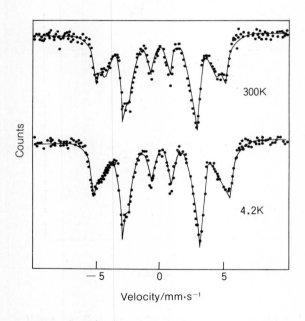

Fig. 4.3.4 ^{57}Fe Mössbauer absorption spectra at 4.2K and
300K for a surface-selectively enriched sample to study the
Fe/Mg interface. The solid curves are the best fit results
which assume superposition of two hyperfine fields. The
hyperfine fields (and the relative intensities) are
330kOe(36%) and 290kOe(64%) at 4.2K and 320kOe(34%) and
270kOe(66%) at 300K. For the smaller hyperfine field
fractions, broader line widths are assumed (30).

surface-selectively enriched sample which was prepared by the successive deposition of ^{56}Fe(70Å), ^{57}Fe(3.5Å) and Mg(970Å) on a mylar substrate. The distribution of the hyperfine field observed here is much narrower than that of the Fe/V interface and the spectra can be interpreted as a superposition of two six-line patterns; one has nearly the same hyperfine field as the standard value of bulk α-Fe and the other, a slightly smaller one. At 4.2K, the values of the two hyperfine fields are 330 and 290kOe, respectively, with an intensity ratio of 36:64. Since 64% of 3.5Å corresponds to one monolayer, the smaller hyperfine field is attributed to the top interface layer. In other words, the hyperfine field of the top layer is smaller by 14% than the bulk, but the second layer already has nearly the bulk value. As for the intensity ratio of the six lines, 3:4:1:1:4:3 is a good assumption, which means the spin direction at the interface is in the plane, similar to the case of the Fe/V interface. The above results indicate that intermixing at the Fe/Mg interface is less than that at Fe/V, which is reasonable since the affinity between Fe and Mg is much weaker than that between Fe and V. Thus it is suggested that a chemically sharp interface may be prepared by combining insoluble elements rather than similar elements.

In these two cases, the top Fe atom layer shows a reduced hyperfine field and a decrease of the magnetic moment is suggested. No fraction with an increased hyperfine field is observed at all. By a similar method, Fe surfaces coated by Ag layers were investigated by Walker's group(31). Ag and Fe are also insoluble with each other. They found not a decrease but a slight increase of the hyperfine field (about 2%) at 4.2K. The technique of using surface-selectively enriched samples can be applied not only to metal/metal interfaces but also to metal/non-metal ones. Rather large increases of the hyperfine field are observed when Fe metal surfaces are coated by such stable ionic materials as MgF_2 or MgO(32). The increase of the hyperfine field of a MgF_2-coated surface is about 13% on the average, and that of a MgO-coating, 8%. Thus the surface (or interface) effect on the magnetic moment is not simple but depends greatly on the coating material.

According to recent theoretical calculations, there seems to be an increase of magnetization at the surface of bulk Fe crystal at zero temperature and also at finite temperatures(14, 17). In these calculations, clean vacuum surfaces, which have the simplest structure, are assumed. On the other hand, in real interfaces, chemical bondings which exist between Fe and the non-magnetic material can reduce the magnetic moment. The degree of reduction depends on the strength of bonding. If two effects are competing at the interface, the magnetization may change in any direction.

Although the local magnetic moment cannot be estimated straightforwardly from the value of the hyperfine field, at least qualitatively the following conclusion may be reached: Fe atoms at the top layer of a bcc(110) surface of Fe crystal show a small increase or a partial decrease of magnetization, depending on the coating material, but no magnetic dead layer exists at all. A magnetically dead state is realized only if a compound layer is formed at the interface. An example of such a case is the Fe/Sb interface which is described later in this chapter.

For further investigation of interface properties, samples should be prepared from appropriate combinations to obtain interfaces with simple and well defined structures. If successfully prepared, samples for interface studies are much easier to handle than those for vacuum surface studies and various experimental techniques can be applied. If the structure is well defined, a computer simulation is possible as in the case of vacuum surface studies and the outcome can be compared with the experimental results. The requirements for preparing an ideal interface sample are to avoid intermixing completely in the case of an epitaxial combination and to identify the local atomic configuration in the case of a non-epitaxial one. Although these requirements cannot be easily met, it is worthwhile to try to prepare samples with well-defined interfaces as close as possible to the ideal.

Not only metallic interfaces but also oxidized surfaces can be studied by Mössbauer spectroscopic measurements on surface-selectively enriched samples (Case-C, Fig.4.3.2)(33). This technique is especially useful for investigating the initial state of oxidation when the oxide layer is very thin. In Fig.4.3.5 the Mössbauer spectra are shown for samples with different ^{57}Fe layer thicknesses after exposure to air. When the ^{57}Fe layer thickness is 80Å, the signal from the oxide part is quite weak since the oxide layer is very thin and most of the ^{57}Fe atoms are in the metallic state. In the case of 20Å, the absorption by the oxide part becomes comparable to that by the metallic part. From the intensity ratio of the two phases at 4.2K, the oxide layer thickness is estimated to be 12Å (converted value to the metal form). Up to this thickness, oxidation proceeds rapidly but further oxidation is rather slow as long as the sample is stored in a dry atmosphere. When the ^{57}Fe layer is only 4Å thick, all ^{57}Fe atoms are in the oxide layer. The hyperfine field at 4.2K is 500kOe on the average, which is close to the value of γ-Fe$_2$O$_3$. Even when an external field of 45kOe is applied, which is enough for the magnetic saturation of normal γ-Fe$_2$O$_3$, the hyperfine fields are not oriented perfectly to the direction of the external field. This result suggests that the spin structure of this oxide layer is speromagnetic, or surface magnetic spins have been canted, as claimed in fine particle studies(22). The spectrum at room temperature also reveals unusual magnetic properties of the surface oxide layer. The pattern is very broad and does not correspond to any compound known in the Fe-O system. The line shape can be explained by the distribution of the hyperfine field and

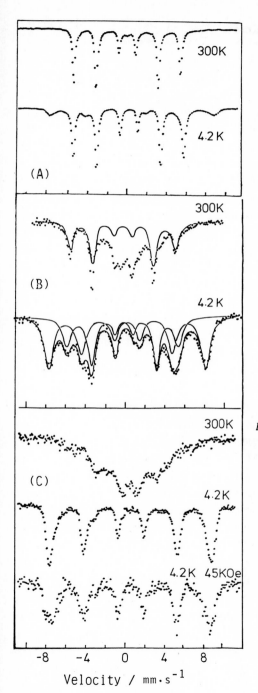

Fig.4.3.5 Mössbauer absorption spectra for surface-selectively enriched samples after exposure to air
(A) The ^{57}Fe layer thickness is 80Å.
(B) The ^{57}Fe layer thickness is 20Å. Solid lines are drawn assuming Lorentzian line shapes.
(C) The ^{57}Fe layer thickness is 4Å. The result in the presence of an external field, 45kOe, is also shown (33).

also by a superparamagnetic relaxation.   Probably both of these mechanisms
coexist and a unique solution cannot be obtained.   It should be mentioned that
the broad line profile at room temperature is fairly reproducible.   In Fig.
4.3.5(B), a very similar broad spectrum is seen in the oxide part of the
$^{57}Fe(20\mathring{A})$ sample.   From the result in Fig.4.3.5(A) for $^{57}Fe(80\mathring{A})$, the oxide part
can be easily identified in the spectrum at a low temperature but is rather
difficult to distinguish in the spectrum at room temperature when the amount of
the oxide part is relatively minor in the sample.

　　The temperature dependence of magnetization at a surface site should be
different from the bulk curve and the deviation is expected to be larger at
higher temperatures.   The temperature dependence of the hyperfine field is
proportional to that of the local magnetization and therefore surface-
selectively enriched samples clearly show the temperature dependence of surface
magnetization.   Unfortunately, since the Curie temperature of metal Fe ($770^{o}C$)
is very high, the deviation from the bulk curve is not noticeable below room
temperature.   However, in the cases of fine particles and thin films, the
deviation can be observed.   The Mössbauer results for thin Fe layers sandwiched
in between MgF_2 layers are shown in Fig.4.3.6(34).   When the thickness is $60\mathring{A}$,
no anomaly is present.   For thinner Fe layers, the amount of the surface
fraction becomes relatively larger and the distribution of the hyperfine field
is observed.   At 4.2K, the surface hyperfine field is apparently larger than the
standard value.   In contrast, the deviation at room temperature is just the
opposite.   The hyperfine field of the surface fraction is smaller than the bulk.
In other words, the temperature dependence of surface magnetization is steeper
than the bulk curve and this tendency is enhanced in very thin films.   A
qualitatively similar result was reported by Walker's group;   in the case of
very thin Fe layers (e.g., less than $35\mathring{A}$), the hyperfine field at 4.2K is larger
than the bulk value by a few %, while that at 300K is smaller also by a few
%(31).   The temperature dependence of surface magnetization is not significant
at low temperatures but becomes unusual when combined with a size effect.

　　Surface magnetic properties of non-metallic materials can also be studied
by using surface-selectively enriched samples(36).   Figure 4.3.7 shows a result
on antiferromagnetic $\alpha-Fe_2O_3$.   Particles of $\alpha-Fe_2O_3$ with a diameter of around
0.1 μm were prepared from ^{56}Fe and then the surface was coated with a monolayer
amount of ^{57}Fe by the wet method.   The observed hyperfine fields are evidently
smaller than the bulk curve.   Although measurement at a temperature higher than
700K is not possible because of the sample's thermal instability, the deviation
from the bulk behavior is apparent.   As shown in the figure, the temperature
dependence of surface magnetization is reproduced when the Weiss field at a
surface site is 60% of that at the interior site.   This treatment is not self-
consistent but is only valid for an isolated impurity site.   However, the

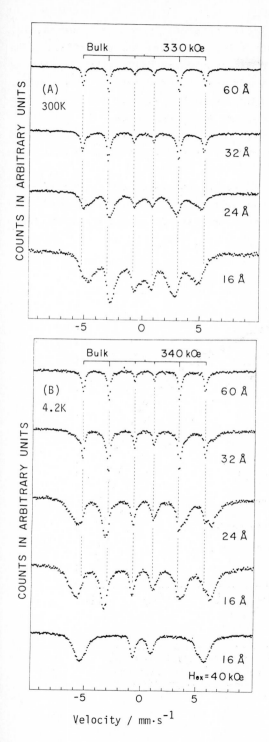

Fig. 4.3.6 Mössbauer absorption spectra for thin Fe films sandwiched in-between MgF$_2$ layers as a function of Fe layer thickness (A) at 300K and (B) 4.2K. The spectrum in the presence of an external field, 45kOe, is also shown, which indicates the Fe layer is fully ferromagnetic(34).

conclusion that the surface Weiss field is reduced seems very reasonable. Similar curves have also been suggested by theoretical treatments(38). The temperature dependence of surface magnetization in Fe_3O_4 estimated from a photoemission experiment is also very similar to the curve in Fig.4.3.7(39). Hence, the curve is regarded as an intrinsic surface behavior and size effect may be neglected. For normal α-Fe_2O_3, the Morin transition, which reflects spin flop, is clearly observed in the Mössbauer patterns. Also at the surface region, the transition is found to occur at the same temperature as that in the interior.

So far there has been no report on the preparation of a [56]Fe base of non-metallic material in the form of film or bulk crystal. Technically it is not difficult to prepare an oxide film sample with a selectively enriched surface in [57]Fe.

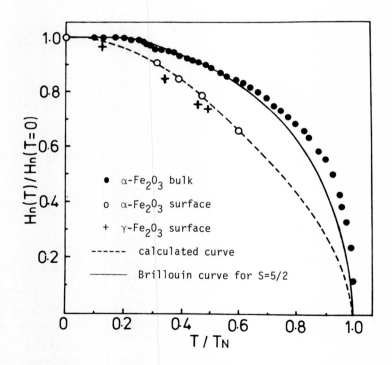

Fig. 4.3.7   Temperature dependence of hyperfine fields in bulk α-Fe_2O_3(35), the surface of α-Fe_2O_3(36) and the surface of γ-Fe_2O_3(37). The dashed curve was calculated with the assumption that the surface exchange field is reduced to 0.6 of the bulk value.

4.4 SUPERLATTICE STUDIES

Enrichment of the isotope is usually not necessary for Mössbauer spectroscopic studies of superlattices. If there are several hundred bilayers and around 50 sample pieces are stacked up, superlattice samples prepared from natural Fe can be used as the Mössbauer absorbers, as far as the substrate is transparent to the gamma-ray. All the samples of Fe/V, Fe/Mg and Fe/Sb superlattice films, whose results from Mössbauer spectroscopic studies are introduced here, were prepared from natural Fe by deposition on polymer substrates. These three combinations are typical examples of an epitaxial superlattice from similar elements, a non-epitaxial one from dissimilar elements and a combination which can form intermetallic compounds, respectively. From the viewpoint of magnetism, all three are ferromagnetic/non-magnetic combinations.

(i) Fe/V superlattices(28)

Figure 4.4.1 shows the ^{57}Fe Mössbauer absorption spectra for Fe/V superlattices measured at 4.2K and 300K. In these samples, the V layer thicknesses are 20Å or more, and therefore the interlayer magnetic interaction through a V layer should be negligible. The Fe layers are thinner than 15Å and the magnetic properties of Fe layer are shown as a function of thickness. The spectrum at 4.2K for Fe(15Å) is a ferromagnetic six-line pattern with a 3:4:1:1:4:3 intensity ratio, which indicates spontaneous magnetization in the plane. Besides the six sharp lines from the bulk part, broad absorptions due to the interface part are also observed. The interface hyperfine field is certainly smaller than that for the bulk and the average value agrees fairly well with that estimated from the measurement on a surface-selectively enriched sample, 240kOe. The sharpness of the spectrum from the bulk part is regarded as evidence that the range of intermixing is very limited. Many Mössbauer studies on Fe-V alloy systems have found that a V impurity in Fe metal decreases the ^{57}Fe hyperfine fields at eight nearest-neighbor sites by about 8%. As a result, the spectrum for Fe(96%)-V(4%) alloy, for instance, shows a splitting of the outermost peaks. From these data, we may conclude that in the case of Fe/V superlattices, diffusion of the V atoms into the middle part of the Fe layer does not take place.

The spectra for the Fe(8Å) sample are much broader, suggesting a wide distribution of the hyperfine field. Nevertheless, it is apparent that the sample is entirely magnetic at 4.2K and no paramagnetic fraction is observed. According to the results on Fe-V alloys, the alloy of Fe(20%)-V(80%) is non-magnetic even at 4.2K. Hence, diffusion of Fe into the V layers seems to be negligible. It is interesting that the line profile looks very similar to that

134

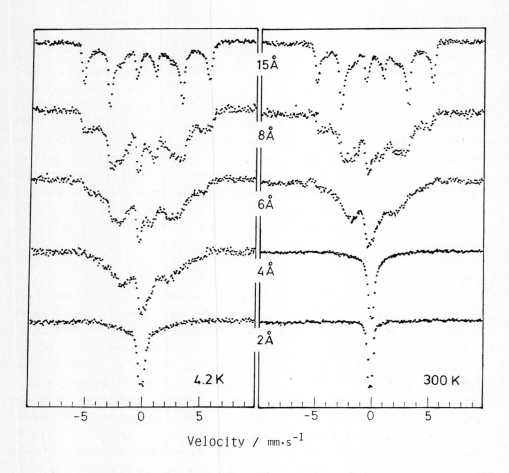

Fig. 4.4.1 Mössbauer absorption spectra at 4.2K and 300K
for Fe/V superlattice film as a function of Fe layer
thickness. V layer thicknesses are 20Å (or more)(28).

of the surface-selectively enriched sample shown in Fig.4.3.3. The thickness,
8Å, roughly corresponds to four atom layers, and if the properties of two layers
are bulk-like and those of the other two are interfacial, the situation should
be similar to that of a surface-selectively enriched sample and the line
profiles may coincide. Thus these observations support the conclusion that the
range of intermixing at the Fe/V interface is about a two-atom distance.
Between 4.2K and 300K, temperature dependence is not significant in the results
for samples thicker than 15Å.

In the spectra for samples thinner than 6Å, temperature dependence becomes noticeable and in addition, a non-magnetic fraction increases with the decrease in the Fe layer thickness. In the case of 2Å, a non-magnetic fraction is dominant even at 4.2K. In general, a magnetic hyperfine structure will disappear under the following conditions: (a) the system is superparamagnetic and the relaxation rate is very fast, (b) each Fe atom has a magnetic moment but the system is not magnetically ordered (the Curie temperature is lower than 4.2K), or (c) each Fe atom has no local magnetic moment at all. In order to clarify the situation, measurement by applying an external field is useful. In Figure 4.4.2, the spectra for an Fe(2Å) sample are shown with and without the application of 45kOe, whose direction is perpendicular to the film plane. A superparamagnetic relaxation can be inhibited by applying 45kOe. Even if the system is purely paramagnetic, 45kOe at 4.2K is strong enough to induce a paramagnetic saturation to a certain degree and then the observed hyperfine field will be different from the external one. On the other hand, the observed hyperfine spectrum in the presence of 45kOe can be simulated by assuming two components of different hyperfine fields, 45 and 15kOe, with a ratio of 73:27, as shown in the figure. In other words, 73% of the total Fe atoms show the same hyperfine field as that applied externally. Therefore, it is concluded that the majority of the Fe atoms in the sample with a nominal thickness of 2Å have no magnetic moment. A minor fraction (the remaining 27%) may hold small magnetic moments.

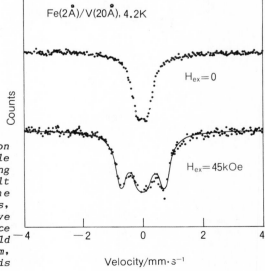

Fig. 4.4.2 Mössbauer absorption spectra for [Fe(2Å)/V(20Å)] sample at 4.2K with and without applying 45kOe. The solid line is the result of fitting by assuming the coexistence of two hyperfine fields, 45 and 15kOe, with a relative intensity ratio of 73:27. Since the direction of the external field is parallel to the gamma-ray beam, an intensity ratio of 3:0:1:1:0:3 is assumed for each component(28).

Even though a nominal monolayer of Fe can be prepared and it has been found to be non-magnetic, we have to be careful in drawing a conclusion. From an analogy of the chemical profile at the Fe/V interface, the synthetic Fe monolayer is regarded as consisting of two layers with a concentration of Fe(50%)-V(50%), rather than one perfect layer with Fe(100%) (see Fig.4.4.3). A theoretical study of the Fe-V alloy system suggests that an Fe atom will lose the magnetic moment in the system when the number of V nearest neighbors is more than 5(40). Fe atoms in a perfect monolayer may hold small magnetic moments since each Fe atom has 4 V neighbors, although this is very close to the critical condition. On the other hand, if the actual structure of the synthetic monolayer is as shown in Fig.4.4.3(B), almost all Fe atoms have 5 V neighbors. And accordingly most of the Fe atoms may lose their magnetic moments. Even though intermixing is only within a few atom layers, the experimental results cannot be directly compared with the theoretical considerations from ideal interfaces. When an interface is formed by combining two similar elements, such a situation is rather common. At any rate, Fe/V superlattices are not suitable as a model system for the purpose of studying a ferromagnetic monolayer.

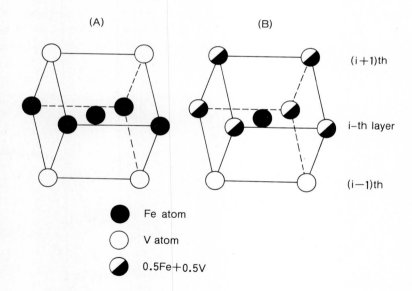

(A)   (B)

(i+1)th

i-th layer

(i−1)th

● Fe atom

○ V atom

◐ 0.5Fe+0.5V

Fig. 4.4.3 Schematic illustration of nearest neighbor atomic configuration. White and black spheres represent Fe and V atoms, respectively. White/black spheres mean 50% occupation.
(A) Ideal structure of Fe monolayer in bcc (110) plane.
(B) Actual structure of a monolayer.

(ii) <u>Fe/Mg superlattices</u>(30,41)

Although Mg is also a non-magnetic metal element like V, the properties of Fe/Mg superlattices are more or less different from those of Fe/V. Even when the thickness is reduced to a monolayer one, each Fe layer exhibits ferromagnetic properties at low temperatures. Figure 4.4.4 shows the Mössbauer spectra of Fe/Mg superlattices as a function of the Fe layer thickness. Here, Mg layer thicknesses are more than 16Å, a thickness which seems to be enough to magnetically insulate each Fe layer. The spectrum for the Fe(15Å) sample is entirely ferromagnetic at both 4.2K and 300K. The intensity ratio is almost exactly 3:4:1:1:4:3 due to the in-plane magnetization. The line shape is well reproduced by assuming the superposition of two hyperfine fields, the values of which at 4.2K are 335 and 315kOe, with a ratio of 65:35(Fig.4.4.5). Since 35% of 15Å corresponds to 5.3Å, which is roughly a two-atom layer thickness, the smaller value, 315kOe, is regarded as the average of the hyperfine fields at the top interface layer. This value is somewhat larger than the interface hyperfine field, 290kOe, estimated from the spectrum for a surface-selectively enriched sample, but the difference is not very large. Apparently the hyperfine fields at the interface region with Mg are larger than those at Fe/V interfaces. The magnetic moment of the Fe atom should be similar to the bulk value even at the top interface layer. In other words, the electronic structure of an Fe atom is not greatly modified by the contact with Mg atoms and this may be due to the weak affinity between Fe and Mg.

If the Fe layer thickness is less than 8Å, the spectra at room temperature become non-magnetic. According to the X-ray diffraction measurement, Fe layers thicker than 15Å have definite bcc structures but those thinner than 10Å are similar to the amorphous. Corresponding to this critical thickness, the magnetic transition temperature also changes rather abruptly and Fe layers thinner than 10Å show non-magnetic spectra at room temperature. At 4.2K, on the other hand, all Fe atoms are ferromagnetic even when the nominal thickness is 1Å. The non-magnetic fraction is not observed at all. The layer with a thickness of 1Å should be an assembly of fractional monolayers (monolayers with limited areas). The average hyperfine field at 4.2K of the Fe(1Å) sample is 260kOe. Even in fractional monolayers, Fe atoms still possess magnetic moments similar to the bulk value. It is interesting that the difference between the monolayer value and the bulk one is nearly twice the difference between the interface value and the bulk. Another strange feature of the spectra for the monolayer region (1 and 2Å) is the intensity ratio, which greatly deviates from 3:4:1:1:4:3 but approaches 3:0:1:1:0:3. The latter ratio indicates the spontaneous magnetization is oriented to the perpendicular direction to the plane. The ratio of the Fe(4Å) spectrum is intermediate and suggests a random orientation of magnetization. Very recently, Gay and Richter calculated the

138

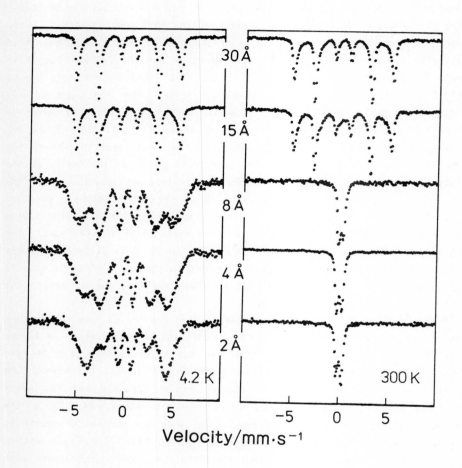

Fig. 4.4.4 Mössbauer absorption spectra at 4.2K and 300K
for Fe/Mg superlattice films as a function of Fe layer
thickness. Mg layers are 16Å (or more)(41).

spin anisotropy of ferromagnetic monolayers of Fe, Ni and V, and the easy
direction of magnetization was suggested to be perpendicular to the plane of the
monolayers of Fe and V, but to be in the plane in the case of the Ni
monolayer(21). The experimental results on Fe/Mg superlattices are thus
consistent with the theoretical prediction.

By evaluating the relative intensities of lines no.2 and no.5, as shown in
Fig.4.4.5, the relative angle of the magnetization on the average is estimated
to be about 70°. A ferromagnetic character is clearly revealed when an external
field is applied(Fig.4.4.6). When 45kOe is applied perpendicular to the film
plane, lines no.2 and no.5 disappear entirely. Then the separation of the

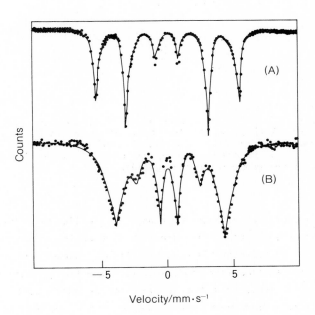

Fig. 4.4.5. Mössbauer absorption spectra at 4.2K for
(A) [Fe(15Å)/Mg(30Å)] and (B)[Fe(1Å)/Mg(16Å)] superlattice
films. The solid lines are the result of fitting. For
the Fe(15Å) spectrum, a superposition of two hyperfine
fields, 337kOe and 313kOe, with a ratio of 65:35 is
assumed. For both 6-line fractions, the intensity ratio
of 3:4:1:1:4:3 is assumed. The Fe(1Å) spectrum is
simulated by six broad Lorentzians whose separation
corresponds to a hyperfine field of 260kOe. The relative
intensity ratio indicates that the average angle between
the magnetization and the film plane is about 70 (30).

140

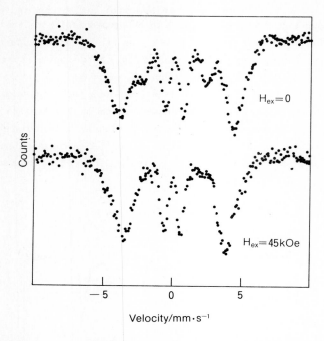

Fig. 4.4.6   Mössbauer absorption spectra for [Fe(2Å)/Mg(16Å)] superlattice film at 4.2K with and without an external field, 45kOe. The direction of the external field is perpendicular to the film plane and parallel to the gamma-ray direction(30).

outermost peaks decreases without any change in the line profiles. If there is a fraction where the electron spin direction is opposite to the external field, it should show an increase of the hyperfine field. Hence it is certain that all electron spins are aligned ferromagnetically. Magnetic transition temperatures are generally estimated from the temperature dependences of the hyperfine fields. The transition temperature decreases with a decrease in the Fe layer thickness and the hyperfine structures collapse at 55K in the case of Fe(2Å) and at 35K in Fe(1Å). The transition temperatures of very thin layers are certainly much lower than the bulk value. However, these temperatures can be regarded as the blocking temperature in a superparamagnetic system and bulk magnetic measurements suggest that short-range spin correlations exist at much higher temperatures.

For the synthesis of a two-dimensional ferromagnetic system, Fe/Mg is a suitable combination. The observed features, perpendicular magnetization and obscure transition temperature, may be regarded as the general characteristics for ferromagnetic monolayers. Unfortunately, the crystal structure of the Fe monolayer is not clear. Even when the Fe layer is a monolayer thick, the structure is not epitaxial to that of Mg. The magnetic moment suggests a structure similar to bcc, which definitely differs from that of fcc or hcp Fe. For further clarification, the structure should be elucidated by measurements such as LEED and EXAFS.

Since Fe and Mg are insoluble with each other, there has been no report even on the properties of Fe impurities in Mg matrix. By depositing two elements simultaneously, a homogeneous alloy can be prepared. Amorphous alloys in a certain compositional range were thus prepared and investigated by van der Kraan and Buschow(42). From the measurements on an Fe(0.5%)–Mg(99.5%) alloy prepared by the co-deposition technique, an Fe impurity dissolved in Mg matrix was found to be non-magnetic(41). On the other hand, the Fe/Mg superlattice samples shown in Fig.4.4.4 are entirely ferromagnetic at 4.2K and non-magnetic absorption is not seen at all. This result indicates that the Mg layers in these samples are chemically pure; that is, their Fe concentrations in Mg layers are almost zero.

(iii) Fe/Sb Superlattices(43)

Although the mutual solubility of Fe and Sb is very limited, intermetallic compounds, FeSb and $FeSb_2$, can be formed. Artificial superstructures are prepared by alternately depositing these elements and the nominal thickness of the Fe layer can be reduced to a monolayer one. The Mössbauer spectra as a function of the Fe layer thickness are shown in Fig.4.4.7. At room temperature, normal ferromagnetic six-line patterns are observed for the samples whose Fe layer is thicker than 15Å. In the spectrum for the Fe(10Å) sample, the magnetic hyperfine structure is fairly broad and a non-magnetic fraction is noticeable. Comparing this result with the data for the Fe/V or Fe/Mg system, we notice that the critical thickness for the Fe layer to hold a stable magnetization at room temperature is always around 10Å.

At 4.2K, the spectrum for Fe(2Å) is almost non-magnetic, which is contrary to the result for Fe/Mg. When an external field is applied, the Fe atoms in the Fe(2Å) sample are found to have no magnetic moment. The relative intensity of this non-magnetic fraction decreases with an increase in the Fe layer thickness. These features also appear in the case of Fe/V superlattices but the difference is that a certain amount of the non-magnetic fraction is always found even in the spectra for layers thicker than 15Å. The amount of the non-magnetic fraction can be estimated from the relative intensity. Then, it is found that the amount roughly corresponds to a thickness of 2Å for each Fe layer, independent of the Fe layer thickness. This result suggests that one atom layer in every Fe layer is magnetically dead.

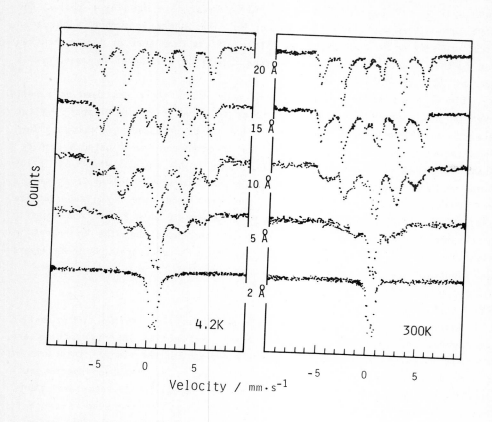

Fig. 4.4.7 Mössbauer absorption spectra at 4.2K and 300K
for Fe/Sb superlattice films as a function of Fe layer
thickness. The thicknesses of Sb layers are 20Å (or
more)(43).

The situation of the Fe/Sb interface can be clarified by using the surface-
selective enrichment technique introduced in the preceding section. Two
samples with different structures are shown in Fig.4.4.8. One was prepared by
depositing in succession Sb, ^{56}Fe(100Å), ^{57}Fe(4Å), and Sb again. In the other
sample, ^{57}Fe(4Å) was deposited before ^{56}Fe. The former sample shows the
situation of the top region of the Fe layer whose nominal thickness is 104Å.
The latter shows the situation where the Fe atom layer is deposited first on the
Sb substrate. Thus from these two samples, we are able to study the "head" and
"tail" interfaces of each Fe layer. As shown in Fig.4.4.8, the spectra for
these samples are quite different.

The former sample, for the study of a head interface, shows a ferromagnetic six-line pattern and the non-magnetic fraction is negligible. Although a structural analysis has not been made for this specific sample, the surface for the most part is regarded as a bcc(110) plane. Undoubtedly, all the Fe atoms up to the top layer are ferromagnetic. The line shape suggests the existence of slightly reduced hyperfine fields, which are due to the interface effect (the effect of Sb-coating). This result is consistent with the observations previously described; that is, the (110) surface of bcc Fe is always ferromagnetic and no dead layer exists. As long as the bcc structure is held, an interface site Fe atom has a magnetic moment which is partially reduced, or enhanced, depending on the material covering the surface.

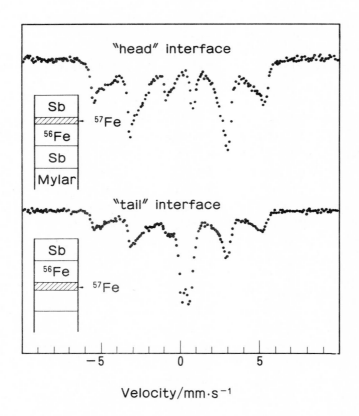

Fig. 4.4.8   Mössbauer absorption spectra at 300K of "head" and "tail" interfaces.   The structures of the samples are illustrated in the figure(43).

On the other hand, the latter sample for the study of a tail interface, shows a significantly large non-magnetic absorption in the spectrum. From the absorption area, the relative amount of non-magnetic absorption is estimated to be about 30% of the total, which corresponds to a nominal thickness of 1.2Å. This result means that about one monolayer of Fe atoms, deposited first on the surface of the Sb layer, has reacted with Sb and subsequently lost the magnetic moment by forming covalent bondings. The magnetic properties of the two interfaces are thus quite different. The Fe atoms at the head interface are ferromagnetic and the chemical profile at the interface seems to change abruptly. At the tail interface, the electronic structure of Fe atoms is greatly modified and one layer is regarded as an intermetallic compound. As the phase diagram suggests, the formation of intermetallic compounds would be easy. Strictly speaking, however, FeSb and $FeSb_2$ are antiferromagnetic at low temperatures and do not correspond exactly to the interface compound whose Fe atoms are non-magnetic.

It is apparent that Fe/Sb superlattices have chemical modulations with unidirectional (not only uniaxial) structures along the direction of film growth. This phenomenon may occur if the chemical reactivity of atom A deposited on substrate B is different from the reactivity of atom B deposited on A. Another possibility is when the microscopic surface roughnesses of layers A and B are greatly different. At any rate, Fig.4.4.8 is the first experimental evidence which clearly shows the difference between the two interfaces. Neutron diffraction and ferromagnetic resonance measurements on Fe/Sb samples also support this model in which one atom layer is magnetically dead(43). This phenomenon may occur in any kind of superlattice. Certainly the difference is very small in the previously described cases of Fe/V or Fe/Mg but might not be zero. On the other hand, Fe/Mn shows a large difference(44).

(iv) Other systems including Fe

Only a few reports have been published concerning Mössbauer spectroscopic studies on superlattices consisting of Fe and elements other than V, Mg and Sb.

Fe/Cu is another example of a ferromagnetic/non-magnetic combination. Superlattices of equiatomic Fe/Cu composition were prepared by van Noort et al. and CEMS measurements were carried out(45). If the Fe layer is thicker than 10Å, ferromagnetic six-line patterns are observed in the spectra, which are similar to the case of Fe/V, and the structure of the Fe layers appears to be bcc. From an analysis of the hyperfine field distribution, the range of intermixing at each interface is estimated to be within three atom layers. Since Fe and Cu are an insoluble combination, the interface may undergo a rather abrupt chemical modulation. In the spectrum for Fe(6Å), a non-magnetic fraction coexists and Fe layers seem to be partially converted to the non-magnetic fcc phase. It is well known that very thin Fe layers epitaxially grown on Cu substrates take the fcc structure. However, in this experiment the observed critical thickness to hold the fcc structure seems to be much less than that reported in previous papers(46). Perhaps this is due to the thickness of the Cu layer, which is also very thin and is the same thickness as the Fe layer. If Fe layers are 10Å thick and Cu layers are thicker than 30Å, the prepared superlattice might be an appropriate sample to study the properties of fcc Fe, or so-called gamma-iron.

The combination of Fe and Mn is ferromagnetic/non-magnetic at room temperature but ferromagnetic/antiferromagnetic at lower temperatures. Because of this feature, the overall magnetic structure therefore might be of particular interest. Even though Fe and Mn occupy nearest-neighbor sites in the periodic table, the structures of Fe/Mn superlattices are very complicated and accordingly the magnetic structures are also not straightforward. Due to the high reactivity of Mn, the interfaces in Fe/Mn superlattices are not chemically sharp. In addition, the two types of interface, head and tail, have very different chemical profiles. The Mössbauer spectra for surface-selectively enriched samples are shown in Fig.4.4.9(44). As in the case of Fe/Sb, a rather

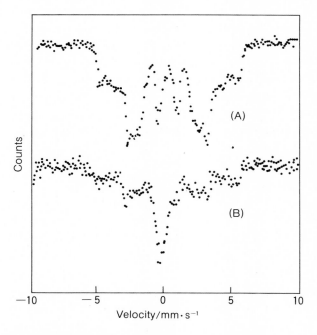

Fig. 4.4.9 Mössbauer absorption spectra at 300K for two kinds of Fe/Mn interfaces(44). The nominal structures are similar to the head and tail interfaces illustrated in Fig. 4.4.8.

abrupt chemical modulation occurs when Mn is deposited on Fe, whereas when Fe is deposited on Mn, interdiffusion over a certain range occurs. The compositional modulation mode in Fe/Mn superlattices is unidirectional along the direction of film growth. The ^{55}Mn NMR technique can be utilized to study Fe/Mn superlattices. In the next chapter, a preliminary result of the chemical profile estimation for Fe-rich Fe/Mn samples (ferromagnetic) is described. For the Mn-rich region, the hyperfine fields of both Fe and Mn are distributed very widely and the discussion is not yet settled. Since the Fe/Mn system is too complicated in many respects, it is not an appropriate choice for preparing a model superlattice sample consisting of ferromagnetic and antiferromagnetic components.

Although the magnetic properties of superlattices consisting of Fe and rare earth metals are of great interest, fundamental as well as technological studies, have been launched only recently and Mössbauer spectroscopic investigations are still in the elementary stage. Umemura et al. examined Fe/Sm(47) and Fe/Gd(48) systems and determined the compositional regions where the magnetization is oriented perpendicular to the plane. Although the vacuum condition was not good and the rare earth layers were supposed to be oxidized, the Mössbauer spectra indicated that the Fe layers are metallic. In the case of UHV deposited Fe/Dy superlattices also, the region for the perpendicular magnetization has been found to be very limited(49). The Fe hyperfine field is decreased by rare earth metal atoms occupying the nearest-neighbor sites. It is not yet clear how the rare earth metal layers contribute to the long-range magnetic order in a ferromagnetic Fe/rare-earth superlattice. For an understanding of the magnetic properties, the results from bulk magnetic measurements and also from neutron diffraction experiments should be combined with the Mössbauer results.

(v) Nuclei other than Fe

Like ^{57}Fe, ^{119}Sn is a very easy nucleus to use as a Mössbauer probe. However, to the author's knowledge, there has been no report on the ^{119}Sn measurement of a superlattice. ^{57}Fe was used to investigate the interface between Fe and Sn and a slow solid state reaction (formation of an intermetallic compound) was detected(50). Since the melting point of Sn is not high, a superlattice with Sn, if prepared, will not be very stable even at room temperature. Sn alloys or compounds are more stable than pure metallic Sn, and thus if the constituent of a superlattice is a Sn compound or alloy, the structure will be stable at room temperature.

In the case of other Mössbauer nuclei, usually not only the measurements but also the data analyses are laborious. ^{121}Sb measurements were carried out on Fe/Sb and Co/Sb superlattices as shown in Fig.4.4.10(51). A wide broadening

in the spectrum for Fe(24Å)/Sb(12Å) indicates a distribution of the hyperfine
fields due to the penetration of spin polarization from the ferromagnetic Fe
layer into the Sb layer. On the other hand, there is almost no broadening for
Co(30Å)/Sb(15Å), in spite of the Co layers being definitely ferromagnetic. The
NMR result of ^{59}Co for Co/Sb superlattices suggests that at each interface Co
atoms have become compound-like and approximately one atom layer is non-
magnetic(see the next chapter). On the other hand, the Mössbauer result on
Fe/Sb described above shows that one layer at the tail interface is magnetically
dead but at the head interface, a ferromagnetic Fe layer is in direct contact
with the Sb layer. The difference in penetrating spin densities of Fe/Sb and
Co/Sb may be attributed to the formation of the compound layer at both head and
tail in Co/Sb but only at tail in Fe/Sb.

While rare earth single crystal superlattices would be interesting subjects
for Mössbauer spectroscopic studies, no measurements have been done yet though
some Mössbauer isotopes are available among rare earth metals.

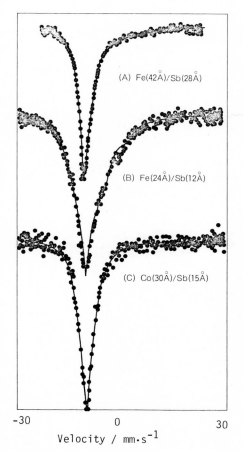

(A) Fe(42Å)/Sb(28Å)

(B) Fe(24Å)/Sb(12Å)

(C) Co(30Å)/Sb(15Å)

-30   0   30

Velocity / mm·s^{-1}

Fig. 4.4.10 ^{121}Sb Mössbauer
absorption spectra at 4.2K for
[Fe(42Å)/Sb(28Å)], [Fe(24Å)/Sb(12Å)]
and [Co(30Å)/Sb(15Å)].
A broadening due to magnetic hyper-
fine interaction is only signif-
icant in the [Fe(24Å)/Sb(12Å)]
spectrum(51).

148

The usefulness of Mössbauer spectroscopy for the study of superlattices may be summarized as follows: The measurements are non-destructive and can be applied to samples with multiphases or inhomogeneous structures. The Fe layer can be closely checked regardless of whether or not contamination, such as oxidation, has occurred or chemical reaction (including formation of an intermetallic compound and interdiffusion) has taken place. The magnetic transition temperatures and the directions of spontaneous magnetization are unambiguously elucidated. The values of the hyperfine field are useful for estimating the magnitudes of local magnetic moments. Finally, from an analysis of the hyperfine field distribution, specific information is obtained on the chemical profile of the artificial modulation.

I wish to acknowledge Y. Endoh, J. M. Friedt, S. Hine, N. Hosoito, K. Kawaguchi, N. Nakayama, K. Takahashi, T. Takada, and R. Yamamoto, who have collaborated in the works described here.

REFERENCES

(1) R. L. Mössbauer, Z. Physik 151 (1958) 124.
(2) See "Mössbauer Effect Reference and Data Journal", which is issued monthly by the Mössbauer Effect Data Center, University of North Carolina, NC. USA. Literature search and copy service are available.
(3) An example of a comprehensive textbook is: N. N. Greenwood and T. C. Gibb, "Mössbauer Spectroscopy", Chapman and Hall, 1971. A recent bibliography is: B. Kolk and Y. Yamada, "Dynamical Properties of Solids, vol.5", G. K. Horton and A. A. Maradudin, eds., North Holland, 1984. A list of review articles is presented therein.
(4) The original idea of the analysis was presented by L. R. Walker, G. K. Wertheim and V. Jaccarino, Phys. Rev. Letters 6 (1961) 98. A book of review articles on isomer shift is: "Mössbauer Isomer Shift", G. K. Shenoy and F. E. Wagner, eds., North Holland, 1978.
(5) R. E. Watson and A. J. Freeman, Phys. Rev. 123 (1961) 2027.
(6) T. Shinjo, J. Phys. Soc. Jpn. 21 (1966) 917.
(7) G. K. Wertheim and J. P. Remeika, Phys. Letters 10 (1964) 14.
(8) V. I. Goldanskii, E. F. Makarov and V. V. Kharapov, Phys. Letters 3 (1963) 334.
(9) M. Blume, Phys. Rev. Letters 14 (1965) 96.
(10) K. Ono and A. Ito, J. Phys. Soc. Jpn. 17 (1962) 1012.
(11) Recent review articles on surface magnetism are: J.C.S. Levy, Surface Science Reports 1(1981)39, G. Allan, ibid 1(1981)121, G. Bayreuther, J. Magn. & Magn. Mater. 38(1983)273.
(12) L. N. Liebermann, D. R. Fredkin and H. B. Shore, Phys. Rev. Letters 22 (1969) 539. L. N. Liebermann, J. Clinton, D. M. Edwards and J. Mathon, Phys. Rev. Letters 25 (1970) 232.
(13) For instance, C. L. Fu and A. J. Freeman, J. Magn. & Magn. Mater. 54-57 (1986) 777.
(14) S. Ohnishi, A. J. Freeman and M. Weinert, Phys. Rev. 28 (1983) 6741.
(15) R. H. Victoria, L. M. Falicov and S. Ishida, Phys. Rev. 30 (1984) 3896.
(16) K. Binder and D. P. Landau, Phys. Rev. Letters 52 (1984) 318. Other references are cited therein
(17) H. Hasegawa, J. Phys. F. Metal Phys. 17(1987)165.
(18) H. Hasegawa, J. Phys. F. Metal Phys. 16(1986)1555.

(19) L. Néel, Compt. Rend. 237 (1953) 1468.
(20) U. Gradmann, Appl. Phys. 3 (1974) 161.
(21) J. G. Gay and R. Richter, Phys. Rev. Letters 56 (1986) 2728.
(22) A. H. Morrish and K. Haneda, J. Magn. & Magn. Mater. 35 (1983) 105.
(23) W. Kiauka, K. Debusmann, W. Keune, R. A. Brand and N. Hosoito, Solid State
     Commun. 58 (1986) 641.
(24) T. Shinjo, T. Matsuzawa, T. Mizutani and T. Takada, Jpn. J. Appl. Phys.
     sup.2, pt.2, (1974) 729.
     T. Shinjo, T. Matsuzawa, T. Takada, S. Nasu and Y. Murakami, J. Phys. Soc.
     Jpn. 32 (1973) 1032.
(25) J. W. Burton and R. P. Godwin, Phys. Rev. 158 (1967) 218.
     Cl. Nourey, G. Machal and Chr. Janot, J. Physique 37 (1976) C6-713.
(26) A. M. van der Kraan, Phys. stat. sol. a18 (1973) 215.
(27) J. Lauer, W. Keune and T. Shinjo, Physica 86-88B (1977) 1407.
(28) N. Hosoito, K. Kawaguchi, T. Shinjo, T. Takada and Y. Endoh, J. Phys. Soc.
     Jpn. 53 (1984) 2659.
(29) N. K. Jaggi, L. H. Schwartz, H. K. Wong and J. B. Ketterson, J. Magn. &
     Magn. Mater. 49 (1985) 1.
(30) T. Shinjo, N. Hosoito, K. Kawaguchi, N. Nakayama, T. Takada and Y. Endoh,
     J. Magn. & Magn. Mater. 54-57 (1986) 737.
(31) A. H. Owens, C. L. Chien and J. C. Walker, J. Physique 40 (1979) C2-74.
     Recently a similar result was reported by J. Korecki and U. Gradmann,
     Hyperfine Interactions 28(1986)931.
(32) S. Hine, T. Shinjo and T. Takada, J. Phys. Soc. Jpn. 47 (1979) 767.
(33) T. Shinjo, T. Iwasaki, T. Shigematsu and T. Takada, Jpn. J. Appl. Phys. 23
     (1984) 283.
(34) T. Shinjo, S. Hine and T. Takada, Proc. 7th Intern. Vac. Congr. and 3rd
     Intern. Conf. Solid Surfaces (Vienna 1977) p.2655.
(35) F. van der Woude, Phys. stat. sol. 17(1966)417.
(36) T. Shinjo, M. Kiyama, N. Sugita, K. Watanabe and T. Takada, J. Magn. &
     Magn. Mater. 35 (1983) 133.
(37) A. Ochi, K. Watanabe, M. Kiyama, T. Shinjo, Y. Bando and T. Takada,
     J. Phys. Soc. Jpn. 50 (1981) 2777.
(38) For instance, T. Takeda and H. Fukuyama, J. Phys. Soc. Jpn. 40 (1976) 925.
(39) S. F. Alvarado, Z. Physik, B33 (1979) 51.
(40) N. Hamada and H. Miwa, Prog. Theor. Phys. 59 (1978) 1045.
     N. Hamada, K. Terakura and A. Yanase, J. Phys. F Met. Phys. 14 (1984) 2317.
     N. Hamada, K. Terakura, K. Takanashi and H. Yasuoka, ibid. 15 (1985) 835.
(41) K. Kawaguchi, R. Yamamoto, N. Hosoito, T. Shinjo and T. Takada, J. Phys.
     Soc. Jpn. 55 (1986) 2375.
(42) A. M. van der Kraan and K. H. J. Buschow, Phys. Rev. B25 (1982) 3311.
(43) T. Shinjo, N. Hosoito, K. Kawaguchi, T. Takada, Y. Endoh, Y. Ajiro and
     J. M. Friedt, J. Phys. Soc. Jpn. 52 (1983) 3154.
(44) T. Shinjo, Hyperfine Interactions 27 (1986) 193.
(45) H. J. G. Draaisma, H. M. V. Noort and F. J. A. den Broeder, Thin Soid Films
     126 (1985) 117. H. M. Noort, F. J. A. den Broeder and H. J. G. Draaisma,
     J. Magn. & Magn. Mater. 51 (1985) 273.
(46) R. Halbauer and U. Gonser, J. Magn. & Magn. Mater. 35 (1983) 55.
(47) S. Umemura, H. Tajika, E. Kita and A. Tasaki, Advances in Ceramics, 16
     (1986) 621.
(48) S. Umemura, H. Tajika, E. Kita and A. Tasaki, IEEE Trans. Mag-21 (1985)
     1942.
(49) T. Shinjo, N. Hosoito, N. Nakayama, K. Yoden and K. Kawaguchi, to be
     published.
(50) T. Shinjo, N. Hosoito, S. Hine and T. Takada, Jpn. J. Appl. Phys. 19 (1980)
     L531.
(51) J. M. Friedt, N. Hosoito, K. Kawaguchi and T. Shinjo, J. Magn. & Magn.
     Mater. 35 (1983) 136. The result on Co/Sb Sample is unpublished.

Chapter 5

NMR Studies on Superlattices

H. Yasuoka
University of Tokyo

5.1 INTRODUCTION

In recent years, considerable efforts have been made in the area of condensed matter physics and chemistry to synthesize new types of materials which are expected to have new physical properties or unique capabilities for future applications. Among them, multilayered films with artificial superstructure have been especially attractive as materials with new artificial periodicity (superlattice). These superlattices, in principle, can even be synthesized from combinations of elements which form neither natural compounds nor solid solutions (alloys); hence we would expect the appearance of new physical properties in terms of structure, magnetism, superconductivity, etc. Moreover, in many cases of metallic superlattices, compositional mixing occurs unavoidably and a topologically two-dimensional alloy region is formed near the interface. Although the degree of mixing depends on the preparation technique and pairing of the two elements, the mixing itself turns out to be of great importance in the following sense: superlattices are formed under a thermally non-equilibrium condition. Therefore, a two-dimensional alloy or compound formed at the interface may have a new structure and/or physical properties which cannot be expected from those in thermal equilibrium.

To study the above two major characteristic features of metallic superlattices, namely, artificial periodicity and the new type of alloy or compound formed at the interface, we have been working on the nuclear magnetic resonance (NMR) in superlattices. NMR is an excellent method for examining the local atomic configuration and the microscopic magnetic properties at the interface region, because the observed quantities from NMR depend directly on the local electronic states. Hence we can obtain microscopic information from the interface region that is distinctly separate from that of the interior region.

Our special interests in metallic superlattices are twofold: one is magnetism, especially new magnetism appearing near the interface and the change in the electronic state under artificial periodicity, and the other is superconductivity. When superconductive layers are separated by non-superconductive layers, the superconductive dimensionality may change with the wavelength of the superlattice. In this chapter, however, we deal only with the magnetic part of NMR studies, focusing on interface magnetism.

It is well known that NMR has contributed much to the understanding of magnetism in general. Although the method of analyzing NMR data is well established by now, we will review briefly in the next section the important aspects which can be extracted from NMR in metallic superlattices. The NMR results for three typical superlattices, [Fe/V], [Co/Sb], and [Fe/Mn], are described in the following sections.

5.2 BASIC ASPECTS OF NMR FOR METALLIC SUPERLATTICES

In NMR experiments of condensed matters, we generally deal with the local magnetic field (H_{loc}) and the electric field gradient (EFG) acting on a nucleus associated with an atom in concern. The former is mainly due to the coupling between the nuclear dipole moment and the net spin density at the nuclear site via the hyperfine interaction. The net spin density arises from the inner core and conduction electron polarization, both of which are proportional to the atomic moment in many magnetic materials. The dipolar field from surrounding atomic moments also contributes to the local field. The EFG contributes to the change in the nuclear energy levels via the nuclear quadrupole interaction with the nuclear quadrupole moment, which is valid for nuclei having a spin $I > 1/2$. This leads to an additional splitting or broadening of the NMR spectrum. Although a microscopic description of the above two basic NMR quantities can be found in many books (1), we review briefly those aspects of NMR relating to metallic (magnetic) superlattices.

Depending on the existence of the local field, the resonance condition of a nucleus in magnetic materials is generally given by,

$$\omega = \gamma_n(H_o + H_{loc}).\tag{5.1}$$

where ω, H_o, and H_{loc} are the angular frequency for resonance, the external magnetic field, and the local magnetic field, respectively. γ_n is the nuclear gyromagnetic ratio which is a constant proper to the nucleus. H_{loc} may be divided into two parts, namely, the static (time averaged) and fluctuating (dynamical) parts.

$$H_{loc} = <H_{loc}> + \delta H_{loc}. \tag{5.2}$$

The static part, $<H_{loc}>$, gives a shift of the resonance position from the position with no local field, and if the system is inhomogeneous, it generally gives the distribution of the resonance spectrum. The fluctuating part of the local field, δH_{loc}, contributes to the nuclear magnetic relaxation which is described later.

In the case of NMR studies of magnetic superlattices, the electronic states, or more specifically the distribution of magnetic moments, should be clarified at each atomic layer. Particularly, it is of great importance to understand how the electronic state varies from layer to layer across the interface and how the atomic configuration is modified near the interface where some degree of interdiffusion may occur between the respective atoms. These features can in principle be studied with NMR as a microscopic probe. For such NMR studies, the practical way to perform the experiment and analyze the data is determined by whether or not we are dealing with atoms in magnetically ordered layers. When the nuclei are situated in magnetically ordered states, NMR can be observed at zero external field and we obtain the spectrum by changing the frequency for resonance. Hereafter this kind of spectrum is referred to as the frequency-spectrum. On the other hand, when the nuclei are situated in the non-magnetic or nearly magnetic sites, that is, $H_{loc} \ll H_o$, the spectrum is obtained at a constant frequency by sweeping the external field. We call this spectrum the field-spectrum hereafter. In the following subsection, we summarize the basic quantities which can be extracted from the NMR experiment for both cases.

5.2.1 NMR in magnetically ordered layers.

Each atom in the magnetically ordered layers has its own saturation magnetic moment or an induced magnetic moment from neighboring ordered moments. In this case, there exists a finite local field acting on the nucleus even without any external field and we can observe the resonance under this field at zero external field. The local field, which we now call the internal field, H_n, is expressed for a homogeneous system as,

$$H_n = \frac{A}{\gamma_n h} <S> = \frac{A}{g\mu_B \gamma_n h} <M_{loc}> \tag{5.3}$$

where A is the hyperfine coupling constant, and $<S>$ and $<M_{loc}>$ are the time averaged electronic spin and the local magnetization, respectively. Since the resonance condition in the present case is $\omega = \gamma_n H_n$, the frequency-spectrum gives us a measure of $<M_{loc}>$. If the system is inhomogeneous such as disordered

154

alloys or the interface of magnetic superlattices, each atom has a different environment depending on the location of the atom concerned. This leads to a distribution of $<M_{loc}>$ and to a characteristic shape of the frequency-spectrum. In order to analyze such a spectrum, we usually use an empirical form in which H_n is assumed to be proportional to the local magnetization, i.e., that of the atom itself and that of its nearest neighbors (n.n.). The field at the i-th position is expressed as,

$$H_n^{(i)} = a\mu_i + b \sum_{n.n.} \mu_{n.n.} \qquad (5.4)$$

μ_i and $\mu_{n.n.}$ are the moments on the i-th and n.n. atoms, respectively, and a and b are to be determined empirically. The first term of Eq.(5.4) is due to the inner-core polarization of the atom itself and the second term is due to the conduction electron polarization by the neighboring moments. Therefore, the distribution of the frequency-spectrum gives the distribution of H_n, hence the distribution of moment around a given atom.

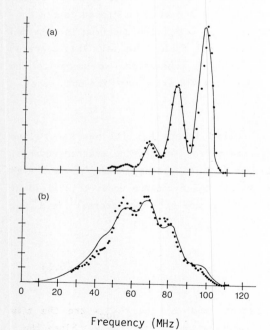

Fig. 5.2.1. ^{51}V spin-echo frequency-spectra in $Fe_{90}V_{10}$ (a) and $Fe_{50}V_{10}$ (b) ferromagnetic alloys obtained at 4.2 K. The full curves show calculated spectra based on Eq. (5.5) in the text.

Frequency (MHz)

A typical example showing the validity of Eq.(5.4) is found in bulk FeV alloy. In Fig.5.2.1, the frequency–spectra obtained by the ^{51}V spin–echo NMR signal are shown for ferromagnetic FeV alloys containing 10% V and 50% V. The spectrum with dilute V concentration shows a main line and distinctly resolved satellite lines, some of which can be seen in Fig.5.2.1 (a). The main line is considered to be associated with V sites having their n.n. shells fully occupied by Fe atoms. The m–th satellite is considered to be associated with V sites having m V atoms and (8-m) Fe atoms in their n.n. shells. (m = 0 corresponds to the main line). For the analysis of the spectrum of 10%V–Fe alloy, we modify Eq.(5.4) as,

$$H_n^i(V) = a\mu_i^V + b_{V-Fe} \sum_{j=1}^{n_{Fe}} \mu_j^{Fe} + b_{V-V} \sum_{j=1}^{8-n_{Fe}} \mu_j^V \qquad (5.5)$$

with $n_{Fe} = (8-m)$,

where μ_i^V is the V magnetic moment itself and $\mu_j^{Fe}(\mu_j^V)$ is the magnetic moment of an Fe (V) atom in its n.n. shell. n_{Fe} is the number of the n.n. Fe atoms. Here, we divide the contribution of the conduction electron polarization into two terms, since the coupling constant between V_i and n.n. Fe, b_{V-Fe}, may have a different value from that between V_i and n.n. V, b_{V-V}. The coupling constant due to the core polarization, a, can usually be fixed for 3d transition metals and alloys, by referring to the theoretical analysis of Akai et.al (2), as

$$a = - 100 \text{ kOe}/\mu_B. \qquad (5.6)$$

Then the coupling constants b_{V-Fe} and b_{V-V} can be determined to explain the H_n of m–th satellite, by using the following assumption of the magnetic moment. From the band structure analysis by Hamada and Miwa (3), the magnetic moment of Fe and V atoms in FeV alloy with dilute concentrations is assumed as follows:

(i) $\mu_j^{Fe} = 2.05 \mu_B$ for all j's, which is the value of the moment of Fe having one V atom in their n.n. shell.

(ii) The V moment is proportional to the sum of the n.n. Fe moments. Hence, we have,

$$\mu_i^V = (n_{Fe}/8) \mu_o^V, \qquad (5.7)$$

where μ_o^V is the V moment in the dilute limit and calculated to be - 0.90 μ_B (3).

Then b_{V-Fe} and b_{V-V} can be determined from H_n for the main ($m = 0$) and the first satellite ($m = 1$) lines as follows.

$$b_{V-Fe} = - 10.8 \text{ kOe}/\mu_B \tag{5.8}$$

and

$$b_{V-V} = - 2.25 \text{ kOe}/\mu_B. \tag{5.9}$$

Once we obtain the coupling constants, the full spectrum can be calculated by assuming a random distribution between Fe and V atoms, where the number of V atoms with a given n.n. configuration obeys the binomial distribution. The intensity of the m-th satellite is then expressed for the bcc lattice as,

$$I_m \propto {}_8C_m \, x^m \, (1 - x)^{8-m}, \tag{5.10}$$

where x is the concentration of V atoms. The calculated spectrum was obtained by convoluting the discrete spectrum with a Gaussian distribution function with an appropriate full width at half maximum, FWHM, and is shown by the solid line in Fig.5.2.1 (a). The FWHM was taken to be 8 kOe here which is considered to be due mainly to the contribution from the atomic configuration beyond the n.n. shell. Using the same parameter, the spectrum for 50%V-Fe alloy is calculated and compared with the experimental spectrum in Fig.5.2.1 (b). The agreement between the calculated and experimental spectra for both cases is reasonably good.

Although the theoretical basis of Eq.(5.4) is rather poor at the moment, the above procedure for analyzing the frequency-spectrum has been quite successful for many transition metal ferromagnetic alloys. Therefore, we have used the same analogy to extract the interface magnetism in magnetic superlattices, as will be shown later.

5.2.2 NMR in non-magnetic or paramagnetic layers.

In this case, the local field is essentially zero unless the external field is applied. Therefore, to observe the NMR signal we usually sweep the external field at a constant frequency and obtain the field-spectrum. The shift of the resonance position due to the local field is defined as the Knight shift, K, and expressed as,

$$K = \frac{H_o - H_{loc}}{H_o} = \frac{A}{\gamma_n h} \frac{\langle S \rangle}{H_o} = \frac{A}{g\mu_B\gamma_n h} \chi_o, \tag{5.11}$$

where χ_0 is the uniform local susceptibility of the atom in concern. Therefore, the distribution of the field-spectrum corresponds to the distribution of the local susceptibility which may depend on the atomic layer in magnetic superlattices. The microscopic origin of H_{loc} is essentially the same as that of H_n, except that it is induced by the external field. However, when we deal with a non-magnetic layer which is sandwiched by the magnetically ordered layers, the main source of H_{loc} is the conduction electron polarization from the ordered layers.

If the interaction of the nuclear quadrupole moment with the electric field gradient is taken into account, the resonance condition, Eq.(5.1), must be modified. When the crystal structure has a uniaxial symmetry (i.e., the electric field gradient has an axial symmetry with no asymmetric parameter) and the electric quadrupole interaction, eqQ interaction, is much smaller than the Zeeman energy, the resonance condition for the transition from the nuclear spin state m to m−1 is given within the first order perturbation as follows.

$$\omega = \gamma_n(H_o + H_{loc}) + \nu_Q(m - 1/2)(3\cos^2\theta - 1), \qquad (5.12)$$

where θ is the angle between the principal axis of the electric field gradient and the direction of the external field, and the average over θ has to be taken in the case of a polycrystalline or powder sample. ν_Q is the coupling constant of the eqQ interaction and is defined as,

$$\nu_Q = \frac{3e^2qQ}{2hI(I+1)}, \qquad (5.13)$$

where Q is the nuclear quadrupole moment and eq is the electric field gradient along the principal axis at the position of nucleus. Equation (5.12) leads to a splitting of the resonance line due to the eqQ interaction which is given in the first order by,

$$\delta\nu_m = \frac{\nu_Q}{2} (m - 1/2)(3\cos^2\theta - 1) \qquad (5.14)$$

Therefore, if the field-spectrum has additional splitting or broadening due to Eq.(5.14), we can estimate ν_Q and from this value determine the charge distribution around the atom in concern is determined. Since the charge distribution is related directly to the atomic configuration, this procedure becomes quite useful for studying atomic arrangement, particularly near the interface where a sizable atomic displacement due to mismatching of the lattice may exist and it creates an eqQ broadening even in cubic materials.

As we already mentioned, the fluctuating part of the local field contributes to the nuclear magnetic relaxation, which is an important observed quantity to obtain information on the dynamical properties of electronic spins. Although the nuclear relaxation is characterized by the transverse and longitudinal decay of nuclear magnetization, we describe only the latter briefly since it is directly related to the electronic dynamics via the nuclear-electron coupling. The longitudinal nuclear relaxation is often called the nuclear spin-lattice relaxation with a characteristic time constant T_1. The relaxation rate $1/T_1$ may be defined in terms of the autocorrelation function of δH as,

$$\frac{1}{T_1} = -\frac{\gamma_n^2}{2} < \delta H_+(t), \, \delta H_-(o) > \exp[i\omega_o t]dt. \tag{5.15}$$

Using the hyperfine field operator H=AS and the fluctuating dissipation theorem, Eq.(5.15) can be rewritten as,

$$\frac{1}{T_1} = -\frac{\gamma_n^2}{2} A^2 < \delta S_+(t), \, \delta S_-(o) > \exp[i\omega_o t]dt$$

$$= 2 \gamma_n^2 A^2 kT \sum_q \frac{\chi''(q, \, \omega_o)}{\omega_o}, \tag{5.16}$$

where k and T are the Boltzmann constant and temperature, respectively. χ'' is the perpendicular component of the imaginary part of wave number, q, and frequency , ω, dependent susceptibility, $\chi(q, \omega)$. Then Eq.(5.16) means that the relaxation rate is proportional to the square of coupling constant, absolute temperature and the q-summation of $\chi''(q, \omega)$ at the resonance frequency of ω_o. Although $\chi(q, \omega)$ depends on the nature of electronic spin dynamics, $1/T_1$ in simple metals, where the effects of electron-electron interaction are neglected, can be related to the density of electronic states at the Fermi surface, $N(E_F)$.

$$\frac{1}{T_1 T} = \sum_k F_k [A^{(k)} \cdot N^{(k)}(E_F)]^2, \tag{5.17}$$

where F_k and $A^{(k)}$ are constant factors and the coupling constants for different electron-nuclear interactions k: d-band core-polarization, d-orbital, conduction electron polarization, etc. $N^{(k)}(E_F)$ is the density of states at Fermi surface for each band. Equation (5.17) simply tells us that if we have the value of A from the Knight shift measurement (see Eq.(5.11)), the measured value of $1/T_1 T$ gives a good estimate of the density of states at Fermi surface which may be modified from that of bulk material by making multilayered film.

So far we have described the basic aspects of NMR in magnetic materials and we see that many fundamental physical quantities can be obtained even from the most primitive NMR experiment. These observed quantities have special features in the sense that they are reflected by quite local electronic states. Therefore, NMR is considered to be an excellent method for the microscopic study of metallic superlattices, since it is expected that there are different electronic states at each atomic layer.

5.3 PROCEDURES FOR NMR EXPERIMENT

In this section, the experimental procedures of NMR for metallic superlattices are summarized. The superlattice sample, $[A/B]_n$, is usually prepared by depositing two elements, A and B, alternately on a substrate of mylar film with temperature T_s by either vacuum deposition or a sputtering technique. For the NMR measurements, a sheet of the deposited film is cut into small sections of about 2.5 cm x 0.7 cm. These sections are stacked and held together by teflon tape, and NMR coil is wound directly around the stack. Since usually 10^{19}–10^{20} atoms per sample volume are needed for the NMR signal to be detected, the deposition times, n, has to be chosen so as to satisfy the above condition at each atomic layer. In order to obtain an NMR signal with an inhomogeneously broadened spectrum, a pulsed NMR (so-called spin–echo) method is usually utilized. Although details regarding technique may vary depending on the researchers, our methods, which are described here, may be classified into three kinds.

a) Measurement of field–spectrum.

A phase–coherent spectrometer is used to obtain the field–spectrum. To observe the spin–echo signal, two rf pulses with time separation τ are supplied by a standard signal generator (SG) with a digitalized DC–pulse generator. The rf pulses are amplified by a wide band amplifier followed by a tuned gated amplifier. The power output of the final amplifier is about 2KW at maximum. The spin–echo signal, which appears at the time τ after the second pulse, is amplified by a tuned low noise amplifier and phase–sensitively detected. The sensitivity of the receiving system is such that the detected signal with a S/N ratio equal to unity corresponds to the input voltage of about 0.3 μV. The detected signals are accumulated by a boxcar integrator. The field spectrum is obtained by recording the integrated spin–echo amplitude with the sweeping of the external magnetic field. The magnetic field is generated either by a conventional electromagnet (H_o < 18 kOe) or superconducting solenoid (H_o < 100 kOe).

b) Measurement of frequency–spectrum.

In order to obtain a reliable frequency–spectrum, a spectrometer which is stable over a wide frequency range is necessary. The exciting rf power is not required to be very high to observe the spin–echo signals associated with the ferromagnetically ordered sites because of the enhancement effect of the applied rf field (4); typically an output power of 5-10 W is adequate. On the other hand, a high-power spectrometer is necessary to observe the signals associated with the antiferromagnetically ordered sites. In this case, we usually utilize a phase-incoherent type rf pulse generator with a power output of about 10 KW and a variable frequency range of 1-1000 MHz. The spin-echo signal is amplified by a low-noise, wide-band amplifier and heterodyned. The IF amplifier is followed by a detector with a diode rectifier. The detected signal with S/N ratio equal to unity corresponds to the input voltage of about 1 μV.

The frequency-spectrum is obtained by plotting the spin-echo amplitude at each frequency point by point. The pulse width is kept at a constant value of more than 1 μsec. at each frequency, so as to obtain a spectrum resolution of less than 1 MHz. To measure the spin-echo amplitude, a pulsed calibration signal is fed externally into the NMR coil and the amplitude of the calibration signal is made equal to that of the spin-echo signal, both being observed on an oscilloscope. The calibration signal voltage is then taken as the amplitude of the nuclear signal. The use of the calibration signal eliminates any change in the sensitivity of the receiver system due to changing frequency and the nonlinearity of the diode detection. The uncertainty of the spin-echo amplitude in the obtained frequency-spectrum is estimated to be less than 10%.

In order to obtain the number of atoms which resonate at a particular frequency, or H_n, from the observed spin-echo amplitude, the following points, some of which apply particularly to NMR in magnetically ordered states, should be heeded. The raw data of the spin-echo amplitudes as a function of frequency, ω, need to be corrected due to: (i) the variation of the transverse relaxation time T_2, (ii) the variation of the Boltzmann factor ($\propto \omega$), (iii) the frequency dependence of the detected voltage induced by the processing magnetization ($\propto \omega$), (iv) the frequency dependence of the enhancement effect of applied rf fields ($\propto \omega$), and (v) the frequency dependence of the enhancement effect of nuclear signals ($\propto \omega$). The spin-echo amplitude depends on T_2 as,

$$E(\tau) = E_o \exp(- 2\tau/T_2), \tag{5.18}$$

Where τ is the separation between the two rf pulses and A is the spin-echo amplitude. In the case that T_2 is different at each frequency, the spin-echo amplitudes extrapolated to $\tau = 0$ (E_o) should be used. The enhancement effect noted in (iv) and (v) above is characteristic of the ferromagnetic sites. However, the frequency dependence noted in (iv) is eliminated when the pulse width is constant at each frequency. Therefore, in the case of signals associated with the ferromagnetic sites, the spin-echo amplitude divided by ω^3(the factors noted in (ii), (iii), and (v) above) is taken to be proportional to the number of nuclei with a given resonance frequency. On the other hand, the spin-echo amplitude is divided by ω^2 in the case of signals associated with the antiferromagnetic sites, because there is little enhancement effect. All the frequency-spectra in our results shown here are thus corrected ones.

c) Measurement of T_1

The spin-lattice relaxation time T_1 is measured by observing the recovery of spin-echo amplitude after the saturating rf comb pulses. The recovery of the nuclear magnetization at the time, t, after the end of the saturation pulses, M(t), is given by,

$$M(t) = M_o[1 - \exp(- \frac{t}{T_1})],$$ (5.19)

where M_o is the nuclear magnetization in thermal equilibrium. M(t) is obtained from the time dependence of the spin-echo intensity, which is averaged digitally in an automatic data taking and processing system, and is best fitted to Eq.(5.19) to get the value of T_1. For magnetic superlattices, we would expect a different value of T_1 at each atomic layer. If this is the case, M(t) has a multi-exponential character and the data should be analyzed with the consideration that spin-lattice relaxation has multi-components.

5.4 NMR STUDIES OF MAGNETIC SUPERLATTICES

Studies of the synthesis and characterization of magnetic superlattices have progressed very rapidly in recent years. In this section, three examples are presented to demonstrate how NMR studies have contributed to the understanding of the microscopic magnetic properties. These examples are (i) [Fe/V] representing a ferromagnetic/paramagnetic superlattice of elements with the same crystal structure (5), (ii) [Co/Sb] representing a ferromagnetic/diamagnetic superlattice of elements with different crystal structures (6) and (iii) [Mn/Fe] representing an antiferromagnetic/ferromagnetic superlattice (7). Before describing the NMR results, we summarize below the basic bulk properties of these mixtures.

(i) Fe-V system: Fe and V are both bcc with lattice parameters, a_{Fe} = 2.86 Å and a_V = 3.03 Å. An Fe-V solid solution can be obtained in all compositions in thermal equilibrium. In particular, the ferromagnetic phase with the CsCl-type ordered structure is stable near the equiatomic composition of Fe and V.

(ii) Co–Sb system: This system cannot form a solid solution in thermal equilibrium but forms intermetallic compounds, CoSb, $CoSb_2$ and $CoSb_3$. The crystal structure of Co metal (usually hcp) is quite different from that of Sb metal (rhombohedral).

(iii) Fe–Mn system: Bulk Mn metal has a very complicated structure with the α–Mn type below 980 K. Although an Fe–Mn solid solution can be obtained in thermal equilibrium like the Fe–V system, the ferromagnetic bcc phase of the FeMn alloy is stable only up to 5 at% Mn. All the other phases are known to be antiferromagnetic (γ –FeMn and α –MnFe).

The primary goal of NMR studies is to find out how the characteristics of magnetism and structure in superlattices, particularly near the interface, are modified from those in thermal equilibrium mentioned above. Below, we describe in some detail the NMR studies for the three types of magnetic superlattices.

5.4.1 NMR studies of [Fe/V] superlattices.

X–ray diffraction measurements of [Fe/V] superlattices have shown that the strongly preferred orientation of [110] of the bcc structure is perpendicular to the film plane (8). By assuming that the diffraction line width arises only from a finite size of crystallite in the superlattice, the grain size is estimated to be about 200 Å and 150 Å perpendicular and parallel to the film plane, respectively. In other words, the superlattice has the $Fe[110]_{bcc}/V[110]_{bcc}$ texture structure (9). A sizable mismatch of each bcc lattice parameter by as much as 6.7% appears to be compensated for by the coherent strains forming near the interface.

In a detailed NMR investigation of $[Fe(15Å)/V(30Å)]_{60}$, the samples were prepared at three different substrate temperatures: (I) T_s = –50 °C, (II) T_s = –30 °C, and (III) T_s = –20 °C. The ^{51}V NMR spin–echo spectra for each sample are shown in Fig.5.4.1 (frequency–spectra) and Fig.5.4.2 (field–spectra); the former was taken at zero external field and 1.3 K and the latter at 14.00005 MHz and 4.2 K.

The frequency–spectra which are associated with the ferromagnetically ordered region near the interface in each sample show a wide distribution of the internal field at the V sites, H_n^V, from about –94 kOe to –23 kOe, although the shape of the spectrum depends strongly on T_s. The negative sign of H_n^V means the direction of H_n^V is antiparallel to that of the magnetization of the Fe layers. This feature was confirmed by the fact that the spectrum shifted to the lower frequency side when the external field was applied. On the other hand, the positions and shapes of the field–spectra for the samples are almost the

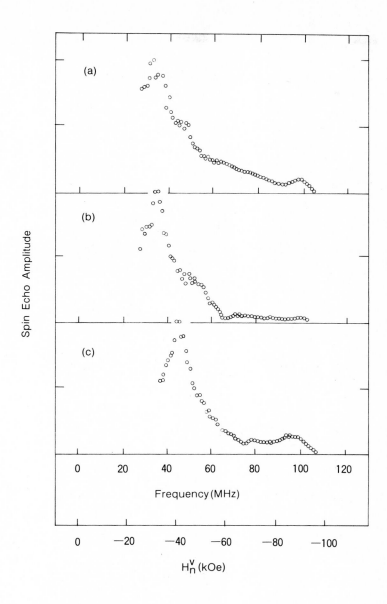

Fig. 5.4.1. ^{51}V *spin-echo frequency-spectra in $[Fe(15\mathring{A})/V(30\mathring{A})]_{60}$ superlattices grown at T_s = -50 °C (a), T_s = -30 °C (b), and T_s = -20 °C (c). All the data were taken at 1.3 K and zero external field.*

164

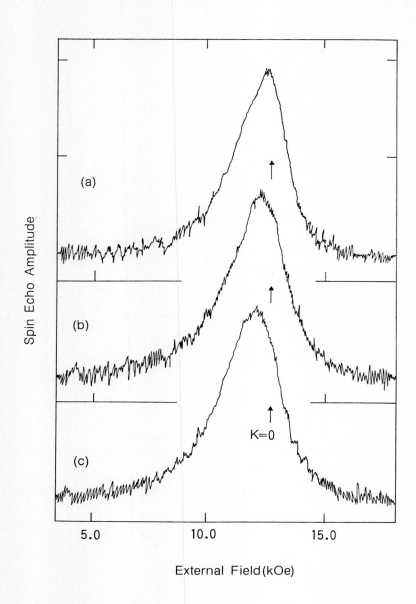

Spin Echo Amplitude

(a)

(b)

K=0

(c)

5.0   10.0   15.0

External Field (kOe)

Fig. 5.4.2. ^{51}V *spin-echo field-spectra in* $[Fe(15\text{Å})/V(30\text{Å})]_{60}$ *superlattices grown at* $T_S = -50$ °C (a), $T_S = -30$ °C (b), *and* $T_S = -20$ °C (c). *All the data were taken at 4.2 K and 14.00005 MHz.*

same.   The field—spectrum has a full width at half maximum, FWHM, of (2.7 ± 0.2)
kOe around the peak at (12.15 ± 0.10) kOe which yields a small positive shift
from the position of zero Knight shift shown by arrow in Fig. 5.4.2 (K =
0 corresponds to H_o = 12.508 kOe at 14.00005 MHz).   The spread of the field—
spectrum is not due to the distribution of the Knight shifts but due to that of
the internal fields, because no change in the width of the spectrum was
observed at different resonance fields.   The distribution of the internal fields
is then estimated to be in the range of −6.7 kOe < H_n^V < 7.5 kOe.   Here it
should be noted that the field—spectra shown in Fig.5.4.2   were obtained only
when the external field was applied parallel to the film plane.   The position
and shape depend on the angle between the direction of the external field and
the film plane.   This anisotropy in the field—spectrum will be shown and
discussed later.

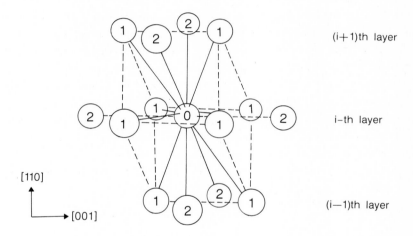

[110]

[001]

*Fig. 5.4.3.   Schematic illustration of the atomic configuration for
the [110] texture of bcc structure.   The atoms denoted as 1 and 2 are
the nearest neighbor and the next nearest neighbor atoms to the atom
denoted as 0, respectively.*

(i+1)th layer

i—th layer

(i—1)th layer

In the frequency—spectra, one can see two main characteristic features,
i.e., there exists a wide distribution of the internal field and this
distribution depends strongly on the substrate temperature.   Since the spectrum
originates from the V atoms which are located near the ferromagnetically ordered
Fe layers, the continuously distributed spectrum indicates that there is some
degree of alloying between the Fe and V atoms near the interface rather than a
sharp separation of the atoms.   If the latter is the case, we should expect to
observe at least discrete lines associated   with the V atoms located at the 1st
(~20 MHz)   and   the 2nd (~15 MHz) atomic layers to the interface (see the

discussion in 5.2.1 and Fig.5.4.3). Therefore, we have assumed that the alloy region due to compositional mixing is formed over a few atomic layers near the interface and the spectrum distribution for three samples has been analyzed.

As is described in 5.2.1, the internal field of ferromagnetic alloys depends strongly on the atomic configuration. The atomic configuration in the interface alloy region may be represented by two physical quantities. One is the chemical composition profile (CCP), that is, the atomic concentration dependence of each atomic layer. The other is the atomic short range order (ASRO). The ASRO parameter α_i in the i-th atomic layer with a concentration of Fe atoms, x_i^{Fe}, may be defined as (10),

$$P(x_i^{Fe}) = x_i^{Fe} + \alpha_i(1 - x_i^{Fe}),\qquad(5.20)$$

where $P(x_i^{Fe})$ is the probability of finding an Fe atom at the n.n. sites of an Fe atom in the i-th atomic layer. $\alpha_i = 0$ means the random configuration and positive (negative) α_i corresponds to a tendency of clustering (ordering). It should be noted that the ASRO between the atoms in different atomic layers is not taken into account here. Then the atomic composition and the ASRO parameter for each atomic layer near the interface have been estimated from the distribution of H_n^V obtained experimentally.

Instead of going into a detailed analysis of the frequency–spectrum which can be found in reference (11), we show here the basic features obtained from a simple treatment of the distribution of the internal field. The H_n^V at given V sites is assumed here simply to be proportional to the sum of the Fe magnetic moments in its n.n. shell, that is,

$$H_n^V = A \sum_{j=1}^{n_{Fe}} \mu_j^{Fe}\qquad(5.21)$$

with A = - 5.3 kOe/μ_B,

where μ_j^{Fe} is the magnetic moment of an Fe atom in the n.n. shell to the V site in concern and n_{Fe} is the number of the n.n. Fe atoms. Equation (5.21) is a simplified form of Eq.(5.5) with the following assumption: (i) the V magnetic moment is proportional to the sum of the n.n. Fe magnetic moment in the whole range of the V concentration and (ii) the third term of Eq.(5.5) is about one order of magnitude smaller than the first and second terms; hence it is neglected. μ_j^{Fe} depends on the atomic configuration surrounding the Fe atom and is determined basically by the number of V atoms in its n.n. shell and the local concentration of Fe atoms (12). Therefore, μ_j^{Fe} can be calculated for a given concentration in each atomic layer with a given ASRO parameter. Equation (5.21) yields a discrete spectrum for $H_n^V(i)$ corresponding to the atomic configuration of each layer which obeys the binomial distribution. The final spectrum is then obtained by convoluting the discrete spectrum with a Gaussian distribution function having an appropriate FWHM. The Gaussian broadening is considered to be due to the contribution of the atomic configuration beyond the n.n. shell. From many trial and error simulations, the best results which reproduce the observed spectra for the three [Fe/V] superlattices are shown in Figs.5.4.4–5.4.6. In these figures, i = 0 refers to the interface alloy layer, and i > 0 and i < 0 correspond respectively to the V and Fe atomic layers (see Fig. 5.4.3). The solid curves in the figures of frequency–spectrum are the best ones calculated using the chemical compositional profile shown in the middle and the ASRO parameter α_i shown in the bottom portion of the figures. It should be noted here that the Fe concentrations are estimated on the assumption that compositional mixing occurs in a symmetrical shape around the interface. Since the treatment here has been greatly simplified, the agreement between the calculated curve and the experimental data points is not very remarkable and there may be an error of about 10% in the values of the parameter. Nevertheless, the following qualitative conclusions may be drawn.

The fact common to all the samples is that only one atomic layer at the interface is a concentrated alloy with an Fe(50%)–V(50%) composition. The compositional mixing is restricted to three or five atomic layers. The chemical composition profile and the ASRO parameters in the interface alloy region vary much from sample to sample. This means that the atomic configuration near the interface depends strongly on T_s. The characteristic results which depend on T_s are summarized as follows:

(i) The ASRO parameters in the interface Fe(50%)–V(50%) layer are positive in the cases of T_s = –50 °C and –30 °C. On the other hand, it is negative in the case of T_s = –20 °C.

(ii) The degree of compositional mixing is small in the case of T_s = –30 °C (three atomic layers) compared with the cases of T_s = –50 °C and –20 °C (five atomic layers).

The ASRO parameters in the interface Fe(50%)–V(50%) layer seem quite likely to be positive for samples grown at T_s = –50 °C and –30 °C. This means that the same kind of atoms tend to cluster in the interface Fe(50%)–V(50%) layer, which is clearly different from the forming of an ordered structure with a negative ASRO parameter in an Fe(50%)–V(50%) alloy in thermal equilibrium.

In the following paragraphs, we discuss the distribution of the Fe magnetic moments. The calculated average Fe magnetic moment in each atomic layer is also shown by dashed horizontal lines in the middle of Figs.5.4.4–

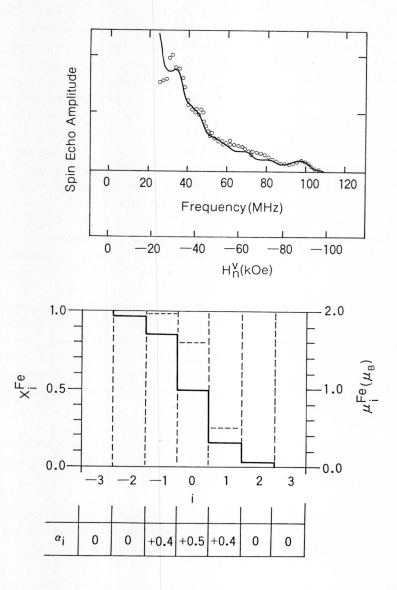

Fig. 5.4.4. *Calculated distribution of $H_n{}^V$, estimated chemical composition profile and ASRO parameters in [Fe(15Å)/V(30Å)] superlattices grown at Ts = -50 °C. The solid curve shown in the upper part shows the best calculated distribution of $H_n{}^V$, which reproduces the experimental data (open circles). The histogram and table in the lower part show the estimated Fe concentration $X_i{}^{Fe}$ (solid line) and the average Fe magnetic moment $\mu_i{}^{Fe}$ (broken line) and the ASRO parameter in each atomic layer near the interface, α_i.*

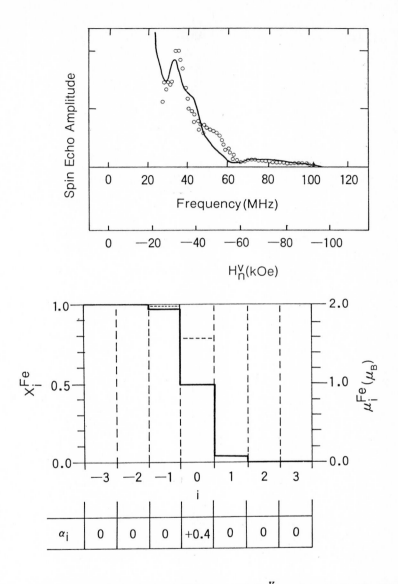

Fig. 5.4.5. *Calculated distribution of $H_n{}^V$, estimated chemical composition profile and ASRO parameters in [Fe(15Å)/V(30Å)] superlattices grown at Ts = -30 °C. The solid curve shown in the upper part shows the best calculated distribution of $H_n{}^V$, which reproduces the experimental data (open circles). The histogram and table in the lower part show the estimated Fe concentration $X_i{}^{Fe}$ (solid line) and the average Fe magnetic moment $\mu_i{}^{Fe}$ (broken line) and the ASRO parameter in each atomic layer near the interface, α_i.*

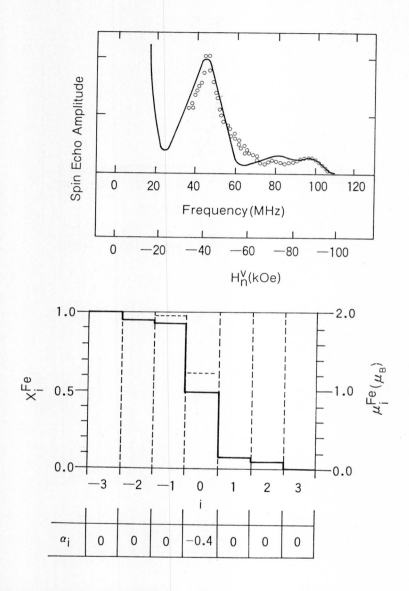

Fig. 5.4.6. *Calculated distribution of* $H_n{}^V$, *estimated chemical composition profile and ASRO parameters in [Fe(15Å)/V(30Å)] superlattices grown at Ts = -20 °C. The solid curve shown in the upper part shows the best calculated distribution of* $H_n{}^V$, *which reproduces the experimental data (open circles). The histogram and table in the lower part show the estimated Fe concentration* $X_i{}^{Fe}$ *(solid line) and the average Fe magnetic moment* $\mu_i{}^{Fe}$ *(broken line) and the ASRO parameter in each atomic layer near the interface,* α_i.

5.4.6.   From the ^{57}Fe Mössbauer measurements of the Fe/V superlattice grown at
T_S = -50 °C, the Fe magnetization at the interface seems to show a reduction of
about 30% compared with the value of bcc Fe metal (13).   The calculated average
Fe magnetic moment at the interface Fe(50%)–V(50%) layer shown in Fig.5.4.3 is
1.61 μ_B, which agrees well with the result from the Mössbauer measurements.

Jaggi et al. also prepared [Fe/V] superlattices independently of our work,
and measured the conversion electron Mössbauer spectra (14).   In their analysis
of the ^{57}Fe internal fields, they adopted a thermal diffusion model and
estimated the diffusion length to be about one atomic layer at the interface.
The chemical composition profile estimated by them is shown in Fig.5.4.7 and
compared with the estimate from our NMR result for a sample grown at T_S = -50
°C.

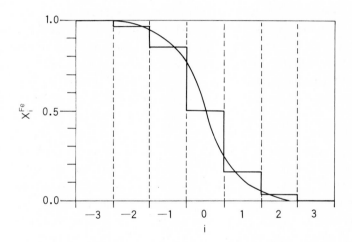

Fig. 5.4.7.   Estimated Fe concentration X_i^{Fe} in each atomic layer
near the interface in the [Fe(15Å)/V(30Å)] superlattice grown at T_S =
-50 °C (step function), which is compared with the chemical
composition profile estimated from the ^{57}Fe Mössbauer spectra by
Jaggi et al. (see Ref. 14).

The agreement is very good, though the preparation techniques were not
quite the same.   However, there is one serious problem, that is whether or not
the compositional mixing is due simply to the thermal diffusion.   It should be
noted that the ASRO parameters are estimated to be positive for samples grown at
T_S = -50 °C and -30 °C.   As mentioned above, this means that the same kind of
atoms in the interface alloy region tend to cluster together, though Fe and V
are known to form a homogeneous solid solution and a negative ASRO parameter is

expected in thermal equilibrium. Therefore the compositional mixing does not seem to be due simply to the thermal diffusion. Considering the origin of the positive ASRO parameter, we make the following speculation about the deposition process at the microscopic level. In many cases, the surface atoms of a layer are likely to cluster because of bonding between them. Let us imagine the following deposition process on a layer having sizable surface roughness, which is now the substrate surface. If T_s is sufficiently low and the kinetic energy of incoming atoms (E_K) is very small, the atomic configuration of the substrate surface may not be disturbed much. Thus we may expect a positive ASRO parameter under the conditions of a low T_s and small E_K, although how low T_s can be and how small E_K can be depend much on the pairing of the two elements. From the above speculation, the compositional mixing in the samples grown at T_s = -50 ^oC and -30 ^oC may be attributed mainly to the microscopic roughness of deposited layers rather than the thermal diffusion. On the other hand, a negative ASRO parameter is observed in the interface alloy region in the sample grown at T_s = -20 ^oC. This clearly indicates that the thermal diffusion toward thermal equilibrium has taken place to some extent near the interface. This is, however, a very surprising result in that the thermal diffusion between Fe and V atoms proceeds radically near the interface even when T_s is increased by only 10 ^oC.

To conclude this discussion on the frequency-spectra, it should be stressed that the atomic configuration near the interface of [Fe/V] superlattices is determined in a delicate balance between the atomic diffusion and the deposition kinetics, both of which depend strongly on a temperature near room temperature. Therefore, a more systematic study of the substrate temperature dependence would contribute to an understanding of the metallurgical problems in metallic superlattices. The annealing effect also provides interesting results of the atomic rearrangement near the interface. These have been left for future NMR studies.

The field-spectra shown in Fig.5.4.2 are considered to be due to the V atoms in the interior region. Although the shape of the spectrum which corresponds to the distribution of the internal field acting on the V atoms has the same characteristic feature for all the samples, it is rather difficult to make a quantitative interpretation. The reason for this is that the internal field is mainly due to the conduction electron polarization by the ferromagnetic Fe layers and we do not have a reasonable model to account for the spatial distribution of the conduction electron polarization in magnetic superlattices. Nevertheless, we have observed the interesting behavior of the field-spectrum which is discussed below.

As mentioned above, the position and shape of the field-spectrum in the [Fe/V] superlattice depend on the angle between the direction of the external

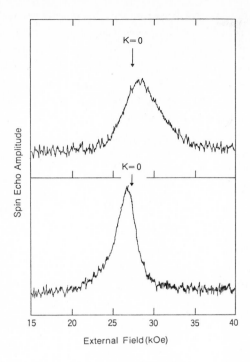

Spin Echo Amplitude

K=0

K=0

15   20   25   30   35   40

External Field (kOe)

Fig. 5.4.8. ^{51}V spin-echo field-spectra in the $[Fe(15\text{Å})/V(30\text{Å})]_{60}$ superlattice obtained at 4.2 K and 30.50005 MHz with the external field parallel (a) and perpendicular (b) to the film plane. The arrows indicate the position of zero Knight shift.

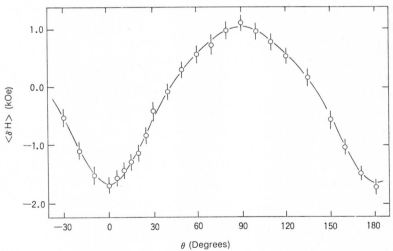

$\langle \delta H \rangle$ (kOe)

1.0

0.0

−1.0

−2.0

−30   0   30   60   90   120   150   180

θ (Degrees)

Fig. 5.4.9. Angular dependence of the shift of the center of gravity from the position of zero Knight shift, $\langle \delta H \rangle$, in the ^{51}V spin-echo field-spectrum in the $[Fe(15\text{Å})/V(30\text{Å})]_{60}$ superlattice. θ is the angle between the direction of the external field and the direction perpendicular to the film plane. The full curve indicates a least squares fit of the data to Eq. (5.22) in the text.

field and the film plane. For example, Fig.5.4.8 shows the field-spectra at 30.500005 MHz with the external field parallel (a) and perpendicular (b) to the film plane. In the case of the perpendicular field, the width of the spectrum is broad and the peak shifts to the negative side of the internal field. For a clearer picture of the angular dependence, this shift of the center of gravity $<\delta H>$, which corresponds to a measure of the mean internal field in the interior V region, is plotted against the angle θ in Fig.5.4.9. Here, θ is taken to be the angle of the external field from the direction perpendicular to the film plane. Since $<\delta H>$ depends on θ with the period of π, the data may be fitted to the following expression,

$$<\delta H> = \sum_n A_n \cos^{2n}\theta. \qquad (5.22)$$

The sufficiently good fitting of the data to Eq.(5.22) was obtained by taking the terms up to n = 3, and the least squares fit yields,

$$A_0 = 1.04 \text{ kOe}, \qquad A_1 = -2.39 \text{ kOe},$$

$$A_2 = 2.43 \text{ kOe} \quad \text{and} \quad A_3 = -2.76 \text{ kOe}. \qquad (5.23)$$

The fitted curve is also shown in Fig.5.4.9.

The anisotropy in the shift may originate from the anisotropic nature of the intra-atomic hyperfine interaction and/or the dipolar interaction with the neighboring magnetic moments. However, from the fact that the coefficients of the $\cos^4\theta$ and $\cos^6\theta$ terms, A_2 and A_3, are quite large and comparable to that of $\cos^2\theta$ term, A_1, and the calculated dipole field amounts to only -100 Oe, it is concluded that the anisotropy of the internal field in the [Fe/V] superlattice is associated with neither the dipolar field nor the anisotropy of the intra-atomic hyperfine coupling constant. The V magnetic moment and/or RKKY-like conduction electron polarization is slightly induced even in the interior V region by the ferromagnetic Fe layers, and the distribution of the internal field depends on the angle between the direction of the Fe magnetic moment and the film plane, probably via spin-orbit coupling. Thus the anisotropy of the internal field may be due to the anisotropy of the distribution of the spin density induced in the interior V region.

The same kind of behavior has been observed in Cu resonance shifts in the [Cu/Nb] superlattice by Yudkowsky et al. (15), although no quantitative interpretation was made. This anisotropic behavior has not been observed in bulk materials and may be very special to magnetic superlattices with a new periodicity. A theoretical study of the anisotropy is anticipated.

The longitudinal nuclear relaxation time, T_1, has been measured in the field-spectrum of [Fe/V] superlattices. The temperature dependence of the rate $1/T_1$ has essentially no angular and sample dependences and is shown in Fig. 5.4.10. $1/T_1$ is almost proportional to temperature, T; that is, the Korringa-like relaxation process is observed. The least squares fit of the data yields $(1/T_1T) = 0.69 \pm 0.01$ $(sec.K)^{-1}$. This value is quite small compared with the value in pure V metal, $(1/T_1T) = 1.26$ $(sec.K)^{-1}$ (16), which is shown by the dashed line in Fig.5.4.10. Although the origin of this reduction is not clear at present, if the reduction is assumed to be due to the reduction of the density of states at the Fermi surface, $D(E_F)$, then $D(E_F)$ in the interior V region in the [Fe/V] superlattice turns out to be reduced by about 25 % compared with $D(E_F)$ in pure V metal. This reduction may be related to the change in the electronic structure, which may be correlated with the fact that the spin density is slightly induced even in the interior V region.

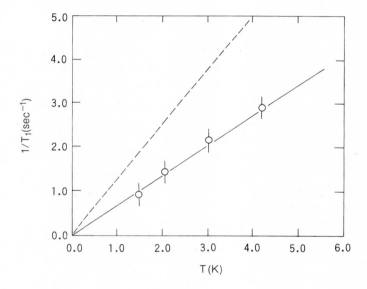

Fig. 5.4.10. Temperature dependence of the spin-lattice relaxation rate $1/T_1$ in the ^{51}V spin-echo field-spectrum of the $[Fe(15Å)/V(30Å)]_{60}$ superlattice. The solid line shows a least squares fit of the data to the form of $T_1T = const.$ (Korringa relation). The broken line shows the temperature dependence of $1/T_1$ in pure V metal $T_1T = 0.79$ (sec.K)).

5.4.2 NMR Studies of [Co/Sb] Superlattices.

X-ray and electron diffraction measurements of [Co/Sb] superlattices have indicated that a coherent structure is not formed between the Co and Sb layers (16).   One layer of each element is considered to be made up of small crystallites with a typical dimension of about 100 Å square in the film plane. When the thickness of the Co layer is larger than 15 Å, the crystal axis of the crystallites in the Co layer is oriented weakly in the direction perpendicular to the film plane, and the preferred orientation is [110] of hcp structure.   The crystallinity of very thin Sb layers is considered to be poor.

The magnetization vs. field curve shows clearly a ferromagnetic behavior for samples with a thickness of the Co layer larger than 15 Å, though the average magnetic moment per Co atom estimated from the saturation magnetization is somewhat smaller than the value of pure hcp or fcc Co metal. On the other hand, for samples with a thickness of the Co layer smaller than 8 Å, the magnetization is remarkably reduced and it does not saturate even when an external field of as much as 50 kOe is applied.   Therefore it is considered that the ferromagnetic character of the Co metal almost disappears in these superlattices.

NMR investigations have been made for ^{59}Co nuclei in four kinds of [Co/Sb] superlattices having different thicknesses of the Co layer but the same Sb layer thickness.   They are [Co(30Å)/Sb(20Å)]$_{100}$, [Co(15Å)/Sb(20Å)]$_{100}$, [Co(8Å)/Sb(20Å)]$_{100}$, and [Co(4Å)/Sb(20Å)]$_{100}$.   The observed NMR signals are classified into two kinds.   Signals associated with the ferromagnetic sites in the interior Co region were observed at 4.2 K in [Co(30Å)/Sb(20Å)] and [Co(15Å)/Sb(20Å)] superlattices but not in [Co(8Å)/Sb(20Å)] and [Co(4Å)/Sb(20Å)] superlattices.   In each frequency–spectrum of the ferromagnetic sites the peak corresponding to pure Co metal was observed.   However, the peak spreads over the resonance frequencies in both fcc and hcp Co metals.   This means that there are many stacking faults in the interior Co region.   Moreover, it should be noted that each frequency–spectrum is extended to the side of a much lower frequency than in pure Co metal.   On the other hand, signals associated with the non-magnetic Co sites have been observed around zero Knight shift for all [Co/Sb] superlattices.   The field–spectrum at 45.0000 MHz for each superlattice is shown as follows in Fig.5.4.11:   (a) [Co(4Å)/Sb(20Å)], (b) [Co(8Å)/Sb(20Å)], (c) [Co(15Å)/Sb(20Å)], and (d) [Co(30Å)/Sb(20Å)].   The integrated intensity of each spectrum is found to be about the same.

The results described above indicate that the Co sites near the interface between Co and Sb layers are non-magnetic over a few angstroms.   Therefore, the ferromagnetic region is considered to disappear in samples with a thickness of the Co layer smaller than about 8 Å.

In the case of the [Co(4Å)/Sb(20Å)] superlattice, the field–spectrum shows

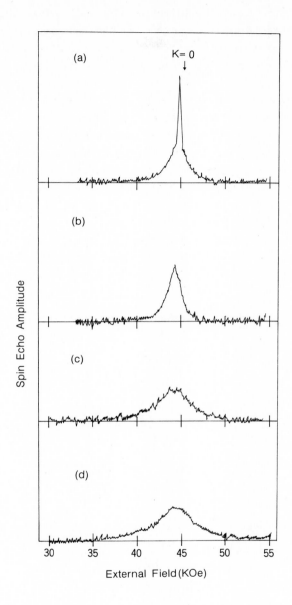

Fig. 5.4.11. ^{59}Co spin-echo field-spectrum associated with the non-magnetic interface regions in $[Co(4Å)/Sb(20Å)]_{100}$ (a), $[Co(8Å)/Sb(20Å)]_{100}$ (b), $[Co(15Å)/Co(30Å)]_{100}$ (c), and $[Co(30Å)/Sb(20Å)]_{100}$ (d) superlattices. All the spectra were taken at 45.00005 MHz and 4.2 K.

a characteristic shape with a sharp central line and a broad skirt. The line
width of the spectrum does not change with the resonance frequency and the
center of the spectrum is shifted from the position of zero Knight shift
proportional to the resonance frequency. Therefore, the characteristic shape of
the spectrum is concluded to be due to the electric quadrupole, eqQ, broadening
rather than the magnetic broadening due to the distribution of Knight shifts or
internal fields. To demonstrate this, a spectrum simulation of the eqQ powder
pattern is shown in Fig.5.4.12 where the amplitude associated with the eqQ
transitions is averaged over the angle between the external field and the
principal axis of the eqQ interaction. From this the coupling constant of the
eqQ interaction, ν_Q, is determined to be 1.8 MHz which is one order of magnitude
larger than that in pure hcp Co metal, 0.2 MHz (17) (ν_Q in pure fcc Co metal is
zero because of the cubic symmetry of the crystal structure). The Knight shift
of the central peak position is found to be 0.7% and independent of
temperature. When the thickness of the Co layer increases, the characteristic
shape due to the eqQ broadening is blurred and the line width increases as shown
in Fig.5.4.11.

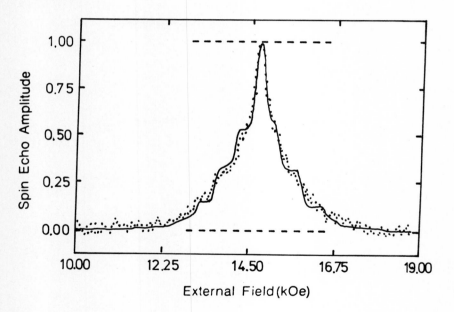

Fig. 5.4.12. ^{59}Co spin-echo field-spectrum in [Co(4Å)/Sb(20Å)] is
compared with the calculated powder spectrum with the eqQ interaction
(solid curve).

The characteristics of the field-spectrum of the [Co(4Å)/Sb(20Å)] superlattice described above may suggest the formation of a compound near the interface in [Co/Sb] superlattices.   As mentioned in the beginning of this section, there are three compounds of Co and Sb in thermal equilibrium, CoSb, $CoSb_2$, and $CoSb_3$.   CoSb is known to be magnetically ordered at low temperature (T < 40 K); therefore the formation of CoSb near the interface may be excluded. $CoSb_2$ is a paramagnetic semiconductor with the arsenopyrite structure, showing a temperature-independent susceptibility. $CoSb_3$ is a diamagnetic semiconductor with the skutterdide structure. When the field-spectrum of the [Co(4Å)/Sb(20Å)] superlattice is compared with that of $CoSb_2$ or $CoSb_3$ (6), it is found that all the spectra have a similar shape and about the same eqQ broadening.   This indicates that the local symmetry of Co atoms in the compound formed near the interface is similar to the structure of $CoSb_2$ or $CoSb_3$.

The above discussion can be extended to samples with a larger thickness of the Co layer.   As the thickness of the Co layer is increased, the original ferromagnetic character of Co metal is considered to develop gradually in the interior Co region.   Hence, the field-spectrum associated with the interface compound region is magnetically broadened by the influence from the interior ferromagnetic Co region.   Thus the typical shape due to the eqQ broadening is smeared out and the line width increases with the increasing thickness of the Co layer.

From the ^{59}Co NMR investigation of [Co/Sb] superlattices, the formation of a compound is suggested near the interface between the Co and Sb layers.   This result is supported by recent X-ray analysis which interprets the relative intensities of small angle diffraction peaks on the assumption of a very thin compound layer near the interface in the [Co(30Å)/Sb(20Å)] superlattice (17).

5.4.3 NMR studies of [Fe/Mn] superlattices

The [Fe/Mn] superlattice is a typical example of a superlattice with alternating ferromagnetic (Fe) and antiferromagnetic (Mn) sublayers.   NMR studies of ^{55}Mn have been made for [Fe(50Å)/Mn(50Å)]$_{80}$, [Fe(50Å)/Mn(15Å)]$_{80}$, and [Fe(50Å)/Mn(8Å)]$_{80}$ superlattices.   X-ray and electron diffraction measurements (18) have revealed that the first combination has a coherent structure consisting of [110] texture of the bcc structure in the Fe layer and [330/441] texture of α-Mn in the Mn layer.   On the other hand, the second and third combinations have no diffraction associated with the α-Mn structure, which suggests that the crystal axis of α-Mn structure in the Mn layer is randomly oriented to different directions both parallel and perpendicular to the film plane, although [110] texture of the bcc structure is formed in the Fe layer. From the magnetization measurement (18), it is known that the magnetization vs. field curve shows a ferromagnetic behavior for all samples.   However, the

180

average magnetic moment per Fe in the second and third types is reduced
substantially compared with that in pure bcc Fe metal and a thermo-remanent
magnetization was observed.   Therefore, [Fe/Mn] superlattices with thin Mn
layers may be mictomagnetic or spin-glass-like, rather than simply
ferromagnetic.

The ^{55}Mn NMR signals in [Fe/Mn] superlattices may be generally classified
into the following two groups: one is associated with antiferromagnetic sites
with the α-Mn structure in the interior Mn region and the other is associated
with the ferromagnetic sites near the interface.   This classification is based
on whether the observed signal has large enhancement effects (ferromagnetic
sites) or not (antiferromagnetic sites).

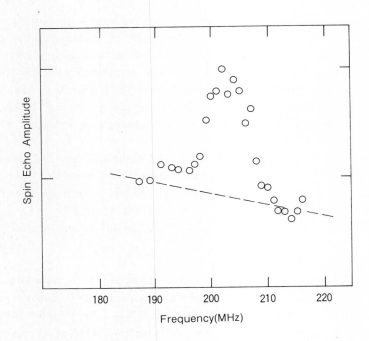

Fig. 5.4.13. ^{55}Mn spin-echo frequency-spectrum associated with site
I in the antiferromagnetic α-Mn structure in the [Fe(50Å)/Mn(50Å)]$_{80}$
superlattice.   The data was taken at 1.3 K and zero external field.
The dashed line indicates the contribution of signals from the
ferromagnetic Mn sites.

The signals associated with the antiferromagnetic α-Mn sites were observed at 1.3 K in the [Fe(50Å)/Mn(50Å)] superlattices, but not in [Fe(50Å)/Mn(15Å)] and [Fe(50Å)/Mn(8Å)] superlattices. This fact is consistent with the result from the X-ray and electron diffraction measurements, that no diffraction lines corresponding to the α-Mn structure were observed in the latter samples. The α-Mn structure is known to have four crystallographically inequivalent sites denoted as I, II, III, and IV (19). However, the frequency-spectra associated with sites II, III, and IV could not be obtained because of the existence of large signals associated with ferromagnetic sites in the same frequency range and/or the lack of sensitivity. Figure 5.4.13 shows the frequency-spectrum associated with site I in the α-Mn structure in the [Fe(50Å)/Mn(50Å)] superlattice, where the background is due to signals associated with the ferromagnetic sites.

It should be noted that the peak of the obtained frequency-spectrum is shifted slightly to the higher frequency side and the line width is somewhat broader than that in pure α-Mn metal. (The frequency-spectrum associated with site I in pure α-Mn metal at 1.4 K has a peak at 199.4 MHz with a FWHM of about 1.5 MHz (20)). The shape of the frequency-spectrum is close to that of α-MnFe alloys with dilute Fe concentrations (21). This fact suggests the existence of a small amount of Fe atoms in the interior Mn region.

On the other hand, signals associated with the ferromagnetic Mn sites were observed in a wide frequency range for all the [Fe/Mn] superlattices. Figure 5.4.14 shows the frequency-spectra at zero external field for [Fe(50Å)/Mn(50Å] (a), [Fe(50Å)/Mn(15Å)] (b), and [Fe(50Å)/Mn(8Å)] (c) superlattices. The sharp peak around 47 MHz is associated with ^{57}Fe nuclei in the Fe layer. This was confirmed from the fact that T_2 of the signal in these frequencies is much longer than that in the other frequencies. All the frequency-spectra shown in Fig.5.4.13 indicate that the distribution of the internal fields at Mn sites, H_n^{Mn}, is extremely wide and the shape of distribution depends strongly on the thickness of the Mn layer. The weighted average of the absolute value of H_n^{Mn} increases with decreasing thickness of the Mn layer. These facts indicate that compositional mixing occurs in a wide range near the interface in the [Fe/Mn] superlattice and the interface alloy region is ferromagnetic even in the Mn dense region. The peak around 242 MHz (H_n^{Mn}=-230 kOe) corresponds to the internal field at Mn sites in bcc FeMn alloys with dilute Mn concentrations (22). This fact suggests again the existence of a small amount of Mn atoms in the interior Fe region.

Although a detailed analysis of the distribution of H_n^{Mn} could not be made since we do not completely understand the crystal structure and the local moment distribution in the ferromagnetic interface alloy region, a crude estimation of the chemical compositional profile has been made from the observed spectra (7).

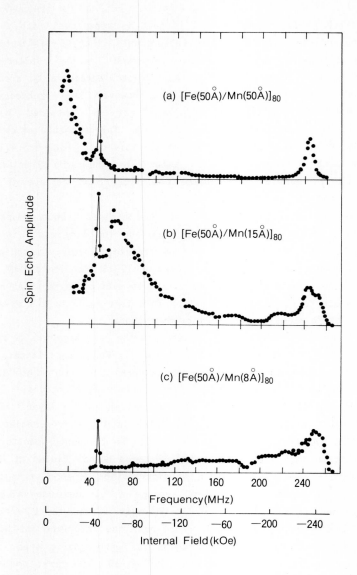

Fig. 5.4.14. ^{55}Mn spin-echo frequency-spectra associated with the ferromagnetic sites in [Fe(50Å)/Mn(50Å)]$_{80}$ (a), [Fe(50Å)/Mn(15Å)]$_{80}$ (b), and [Fe(50Å)/Mn(8Å)]$_{80}$ (c) superlattices. All the data were taken at 1.3 K and zero external field. The sharp peak around 47 MHz is the signal associated with the ^{57}Fe nuclei in the Fe layer.

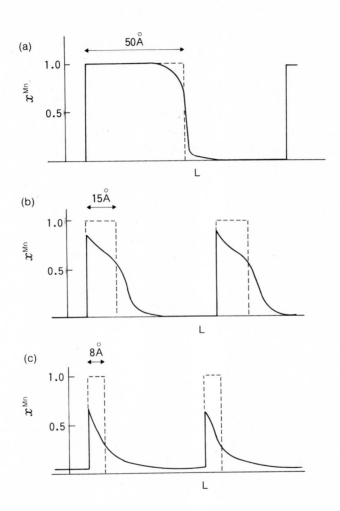

Fig. 5.4.15. Calculated chemical composition profiles of [Fe(50Å)/Mn(50Å)] (a), [Fe(50Å)/Mn(15Å)] (b), and [Fe(50Å)/Mn(8Å)] (c) superlattices (solid curves). X^{Mn} and L indicate the Mn concentration and the distance along the perpendicular direction to the film plane, respectively. The broken lines show the chemical composition profiles for the case of completely flat interfaces.

The results are shown in Fig.5.4.15. In this calculation, the following is assumed: (i) the distribution of Mn and Fe atoms in the ferromagnetic alloy is random; that is, the ASRO is not taken into account and (ii) the compositional mixing occurs only when an Fe overlayer is deposited on a Mn layer substrate. The latter assumption is made from the result of ^{57}Fe Mössbauer measurements, when the internal field at Fe sites in the interface Fe region overcoated with the Mn layer is close to that in pure bcc Fe metal, while the internal field at Fe sites in the interface Fe region deposited on the Mn layer substrate is remarkably reduced when compared with that in pure bcc Fe metal (23).

While the calculated compositional profile shown in Fig.5.4.15 is rather crude, it is nevertheless observed that formation of the alloy region near the interface of [Fe/Mn] superlattices is extended in much wider atomic layers than in the case of [Fe/V] superlattices (at least 15 Å in the former). The most striking feature is that the alloy is ferromagnetic even in the layers with dense Mn concentration in contrast to the fact that alloys in thermal equilibrium are mostly antiferromagnetic, except for very dilute Mn concentration. In this regard, the alloy formed near the interface may be concluded to be a new type of ferromagnetic FeMn alloy, the structure of which is not clear at the moment, but it may be inferred that the non-equilibrium bcc phase is stabilized over the dense Mn region.

5.5. CONCLUDING REMARKS

The NMR investigations of magnetic superlattices have enabled us to explore microscopically the structural and magnetic properties of the interface alloy region. In this chapter, we have presented the results for three typical cases. The characteristic features emerging from the NMR investigation may be summarized as follows.

In [Fe/V] superlattices, it has been found that only one atomic layer at the interface is a concentrated alloy with the Fe(50%)-V(50%) composition. Compositional mixing is restricted to three or five atomic layers. Therefore the interface in the [Fe/V] superlattices is fairly sharp compared with that in [Ni/Cu] superlattices reported by Gyorgy et al.(24), even though the metallurgical properties of bulk FeV and NiCu systems are similar to each other. The ASRO parameter of the interface alloy region in the [Fe/V] superlattices is estimated to be positive in samples grown at $T_s < -30$ °C. This is a very suggestive result, because the formation of FeV alloy with a positive ASRO parameter cannot be expected in thermal equilibrium. In [Fe/Mn] superlattices, compositional mixing is considered to occur over a wide range (at least 15Å) near the interface. The most important point is that the interface alloy is ferromagnetic even in the Mn dense region, which cannot be expected in thermal equilibrium. Thus it is concluded that new types of alloys are created in [Fe/V] and [Fe/Mn] superlattices from the multilayers. On the other hand, compound formation near the interface in [Co/Sb] superlattices is suggested, where the local symmetry of Co atoms may be similar in nature to $CoSb_2$ or $CoSb_3$. This result seems to correspond to the eutectic type of phase diagram between Co and Sb, in comparison with FeV and FeMn systems which can form solid solutions.

Finally the origin of compositional mixing near the interface should be mentioned. We may consider two possibilities, the microscopic roughness of deposited layers and the thermal diffusion. Which factor is predominant seems to depend strongly on T_s, E_K (the kinetic energy of deposited atoms), the deposition rate, and the pairing of the two elements in superlattices. For example, in the [Fe/V] superlattice, the positive ASRO parameter of the interface alloy region seems to indicate that thermal diffusion is not very effective at $T_s < -30$ ^oC. On the other hand, compound formation in the [Co/Sb] superlattice means that a rearrangement of Co and Sb atoms due to the thermal diffusion occurs near the interface at $T_s = -50^oC$. The process of the formation of the ferromagnetic alloy region near the interface in the Fe/Mn superlattice is not clear only from the present study. In the future, the elucidation of the microscopic mechanism of the compositional mixing near the interface in relation to the kinetics of the deposited atoms on the substrate surface should be a major goal in the study of artificial metallic superlattices. As demonstrated in this chapter, NMR can be used widely, we believe, as a sensitive tool for the study of the above problems.

The author wishes to acknowledge T. Shinjo, K. Takanashi, and N. Hosoito, who have collaborated in the works described here.

REFERENCES

(1) For example, I. D. Weisman, Techniques of Metal Research, Vol. VI, Part 2, ed. R. F. Bunshah, (John Wiley & Sons, New York, 1973).
(2) H. Akai, M. Akai and J. Kanamori, J. Phys. Soc. Jpn., 54 (1985) 4257.
(3) N. Hamaka and H. Miwa, Prog. Theor. Phys., 59 (1978) 1045.
(4) A. M. Portis and A. C. Gossard, J. Appl. Phys., 31 (1960) 205.
(5) K. Takanashi, H. Yasuoka, K. Kawaguchi, N. Hosoito and T. Shinjo, J. Phys. Soc. Jpn., 53 (1984) 4315.
(6) K. Takanashi, H. Yasuoka, K. Takahashi, N. Hosoito, T. Shinjo and T. Takada, J. Phys. Soc. Jpn., 53 (1984) 2445.
(7) K. Takanashi, H. Yasuoka, N. Nakayama, T. Katamoto and T. Shinjo, J. Phys. Soc. Jpn., 55 (1986) 2357.
(8) N. Hosoito, K. Kawaguchi, T. Shinjo, T. Takada and Y. Endoh, J. Phys.Soc. Jpn., 53 (1984) 2659.

186

(9) Y. Endoh, K. Kawaguchi, N. Hosoito, T. Shinjo, T. Takada, Y. Fujii and T. Ohnishi, J. Phys. Soc. Jpn., 53 (1984) 3481.
(10) J. M. Cowley, Phys. Rev., 77 (1950) 669.
(11) N. Hamada, K. Terakura, K. Takanashi and H. Yasuoka, J. Phys. F (Metal Physics), 15 (1985) 835.
(12) T. Asada, K. Terakura and T. Jarlborg, J. Phys. F (Metal Physics), 11 (1981) 1847.
(13) N. Hamada, K, Terakura and A. Yanase, J. Phys. F (Metal Physics), 14 (1984) 2371.
(14) N. K. Jaggi, L. H. Schwartz, H. K. Wong and J. B. Ketterson, J. Mag. & Mag. Matter., 49 (1985) 1.
(15) M. Yudkowsky, W. P. Halperin and Ivan K. Schuller, Phys. Rev., B31 (1985) 1637.
(16) N. Nakayama, K. Takahashi, T. Shinjo, T. Takada and H. Ichinose, Jpn. J. Appl. Phys., 25 (1986) 552
(17) M. Kawakami, T. Hihara, Y. Koi and T. Wakiyama, J. Phys. Soc. Jpn., 33 (1972) 1591
(18) N. Nakayama, T. Katamoto, N. Hosoito, T. Shinjo and T. Takada, to be published.
(19) T. Yamada, J. Phys. Soc. Jpn., 28 (1970) 596.
(20) H. Yamagata and K. Asayama, J. Phys. Soc. Jpn., 33 (1972) 400.
(21) T. Kohara and K. Asayama, J. Phys. Soc. Jpn., 37 (1974) 393.
(22) M. Rubinstein, G. H. Stauss and J. Dweck. Phys. Rev. Lett., 17 (1966) 1001.
(23) T. Shinjo, Hyperfine Interactions 27 (1986) 193.
     N. Nakayama, T. Katamoto and T. Shinjo, in preparation.
(24) E. M. Gyorgy, D. B. McWhan, J. F. Dillon Jr., L. R. Walker and J. V. Waszczak, Phys. Rev. B25 (1982) 6739

Chapter 6

Superconductivity in Superlattices

V. MATIJASEVIC and M. R. BEASLEY
Stanford University

6.1 INTRODUCTION

The superconducting materials community pays particular attention to artificially-structured materials. In fact, much of the early work in artificially-structured metals as a whole was concerned with superconducting materials. The reasons are partly historical and partly technical. There is a long tradition of seeking new and novel materials within the field of superconductivity. This tradition seeks to find new superconducting phenomena, to better understand the mechanisms of superconductivity, and of course to find materials with higher superconducting tradition temperatures. The use of physical vapor deposition has played a large role recently in this quest for new superconducting materials. Thus, it is natural that this community would turn its attention to the possibilities of artificially- structured materials. At the same time, the characteristic length scales of superconductivity (e.g., the superconducting coherence length) are relatively long compared with other cooperative phenomena, and therefore it is reasonably easy to layer materials on length scales that will dramatically affect the superconducting properties of a material.

Several reviews of the work in this area have appeared recently(1-3) and give a good account of the main issues that have tended to be the focal points of the field. They also describe the various types of artificially-structured superconductors that have been successfully fabricated and the various vapor deposition techniques that have been used to make these materials. In this review, we stress more recent results both from the materials and physics points of view. In particular we attempt to ascertain the new directions and the outstanding opportunities in the field.

We begin in Section 6.2 with a summary of the various artificially-structured superconductors that have been fabricated to date. These are cataloged according to materials types, the quality of the layering and the dimensionality of the resultant materials. In keeping with the spirit of this review the emphasis is on new entries and on those materials or materials combinations for which the highest quality layering has been achieved. Next, in Section 6.3, we discuss results from recent theoretical calculations. Finally, in Section 6.4 we discuss the open or incompletely answered questions that we see arising from the newer experimental and theoretical work.

6.2 REVIEW OF EXPERIMENTAL RESULTS

Over 30 different systems of synthetically-layered materials exhibiting superconducting properties have been reported in the literature. Table 6.1 shows most of these. The multilayers are listed according to the basic pair of materials used in the layering. This table is an updated version of the listing published in a previous review(1). A number of new systems and some improved samples of previously studied systems have been reported since then. In this newer work the researchers in the field have tended either to look for better quality multilayers, in terms of their structure, or for new phenomena arising by new combinations of materials.

Vapor deposition techniques that are used in multilayer preparation have become rather advanced in the past years. At the same time characterization techniques, X-ray, Auger, TEM, and others, are used more routinely by researchers, giving more detailed information about the degree of order in the multilayer system than was typical in the past. In particular, TEM methods have been developed for examining mass contrast as well as diffraction patterns from the layers(55-57). For these reasons we emphasize characterization of the multilayer order rather than the sample preparation technique. We note, though, that most of the systems in this table were prepared either by e-beam evaporation in an ultra high vacuum or sputter deposition, with the latter increasingly common. In addition special techniques have been developed for composition-modulated structures, such as computer control either for the substrate table(58,59) or for source shutters(47). Literature references for descriptions of particular preparation techniques are given in the last column of the table.

The possibilities for multilayer structures, in terms of materials and layering geometry, are vast and largely unexplored. For example, all of the multilayers in the table are composed of only two different constituent layers. In principle one can compose a structure with an arbitrary number of different layers. Most of these multilayers have, as well, only one thickness for each one of the constituent layers, making the layering periodic. Again, it is possible, in principle, to vary the thicknesses of the layers within the multilayer. Some of the newer work has addressed novel possibilities for layering geometries. Several works have looked at quasiperiodic layering(44,28), where one of the layers has two possible thicknesses, and another work(47) has varied the thickness of one constituent in different ways, including fractal geometries.

Table 6.1 lists multilayer systems made up of elements first, followed by compound and alloy systems. The superconducting layer with a higher transition temperature is listed first using notation A/B, where A and B are the nominal constituent layers. The thicknesses of the two layers, in Angstroms, respectively are given in the next column, and the wavelength Λ of a periodic multilayer is given by the sum of the two layer thicknesses.

It is useful to classify multilayers according to the degree of their chemical and structural order (see Chapter 2 of this volume). The intralayer structural order is given in column three of the table. Only the crystal structure of the individual layers is noted here. Unless specified otherwise the layers are assumed to be polycrystalline and textured. Most researchers have done detailed X-ray analysis as a matter of routine and therefore have obtained much more quantitative as well as more qualitative information about the structure. Since we are concerned here primarily with the superconducting properties of the multilayer system, we have omitted detailed information about the structure that does not affect the superconducting properties in an essential manner.

Chemical order is typically fairly high for most of these systems. However, a certain amount of interdiffusion, even for the structurally most ordered systems, is always present and indeed is often important for the superconducting properties of the material as a whole. In a number of cases the interfacial region is believed to govern the overall behavior. As an example, the fact that the Au/Ge multilayer even superconducts is likely due to diffusion. For most cases, however, it is not the determining factor, but must still be examined.

Nominal types of superconducting multilayer systems are given in the column describing the layer type. The possibilities are: multilayers composed of two different superconducting layers S/S', multilayers in which one of the layers is a normal metal or at least a material that has finite conductivity at low temperatures S/N, systems in which one material is intrinsically magnetic S/M, and systems where one of the layer materials is basically insulating S/I. Where relevant we have placed a presumed interfacial layer in parenthesis. Usually this is important only if that interfacial region is superconducting, and particularly if it has a higher critical temperature than the rest of the material. We should note though that degradation of superconducting material at

TABLE 6.1

System	layer thicknesses (Å)	structure	layer type	dimensionality	sc properties studied	references
Al/Ge	30-250	granular/amorphous	S/I	2D	T_c, H_{c2}	Haywood and Ast (4)
Au/Ge	10/13	-	N/(S)/I	3D	T_c, H_{c2}	Akihama and Okamoto (5)
In/Ag	100	-	S/N	2D	T_c	Granqvist and Claeson (6)
Mo/Ni	7, 8-150	amorphous, bcc/fcc	S/M	2D, quasi-2D	T_c, H_{c2}	Uher et al (7,8,9)
Mo/Sb	13-58/7-63	bcc/hcp	S/N		T_c	Asada and Ogawa (10)
Nb/Al	10-150	bcc/fcc	S/(S")/S'	2D	T_c, tunneling H_{c2}	Geerk et al (11); McWhan et al (12); Guimpel et al (13); Chevrier et al (14)
Nb/Cu	5-10,000	bcc/fcc	S/(S')/N	2D, quasi-2D	T_c, H_{c2}, λ tunneling	Banerjee et al (15,16); Chun et al (17); Yang et al (18); Guimpel et al (19)
Nb/Ge	20-100	bcc/amorphous	S/(S')/I	2D, quasi-2D	T_c, H_{c2} fluctuation conductivity	Ruggiero et al (20,21)
Nb/Pt	-	bcc/fcc	S/N	-	T_c	Karkut el al (22)
Nb/RE (RE=Er,Lu,Tm)	25-1000	bcc/hcp	S/M	-	T_c	Greene et al (23)
Nb/Ta	10-125	bcc/bcc single-crystal	S/(S")/S'	2D, quasi-2D	T_c, H_{c2}, C_p tunneling	Durbin et al (24,25); Hertel et al (26); Broussard et al (27); Lin et al (28)
Nb/Ti	3-3000	bcc/bcc	S/(S')/S'	2D, quasi-2D	T_c, H_{c2}	Zheng et al (29); Qian et al (30)
Nb/Zr		bcc/bcc	S/(S")/S'	2D	T_c, H_{c2}, C_p	Lowe et al (31); Claeson et al (32); Broussard et al (33)
Pb/Bi	300-4000	-	S/N	3D	J_c	Raffy et al (34,35,36)

System	layer thicknesses (Å)	structure	layer type	dimensionality	sc properties studied	references
Pb/Fe	≤300	-	S/M	2D	T_c, tunneling	Cleason (37)
Ru/Ir	10,41,62	alloy,hcp/hcp, hcp/fcc	S/N	-	T_c	Clarke et al (38)
Sn/Ag	100	-	S/N	2D	T_c	Granqvist and Claeson (6)
V/Ag	100-360	bcc/fcc	S/N	2D, quasi-2D	T_c, H_{c2}, λ_{eff}	Kanoda et al (39,40,41)
V/Fe	200-1000/1-15	bcc/bcc	S/M	2D, quasi-2D	T_c, H_{c2}	Wong et al (42)
V/Mo	30-250/15-65 quasiperiodic	bcc/bcc	S/S'	2D, quasi-2D	T_c, H_{c2}	Karkut et al (43,44,45)
V/Ni	5-10,000/4-100	bcc/fcc	S/M	2D, quasi 2D	T_c,H_{c2}	Homma et al (46)
V/Pt	-	bcc/fcc	S/N		T_c	Karkut et al (22)
MoGe/MoGe	9-150/9-1000 fractal	amorphous/amor.	S/S'	2D, quasi-2D	T_c, H_{c2}	Matijasevic and Beasley (47); Missert and Beasley (48)
Nb/CeCu$_6$	20-1000		S/S'	-	T_c	Greene et al (49)
Nb$_3$Ge/Nb$_3$Ir$_3$	750, 1500	A15/A15	S/S'	3D	T_c	Schmidt et al (50)
Nb$_3$Ge/Nb$_3$Ge		A15/A15	S/S'		T_c	Yamamoto et al (51)
NbN/AlN	15-300		S/S'	2D, quasi-2D	T_c, H_{c2}	Murduck et al (52)
Nb$_3$Sn/Y	2500-10,000	A15	S/N	3D	J_c	Howard et al (53)
NbTi/Ge	.5-200/28-500	bcc/amorphous	S/I	2D	T_c, H_{c2}, J_c	Qian et al (30); Jin et al (54)

the interface, although not noted explicitly in our table, is often important, for example, as a limiting factor for T_c behavior versus thickness.

The table also gives the superconducting dimensionality of the layered system. This is determined according to whether the superconducting layers are thick or thin compared with their intrinsic (intralayer) superconducting coherence length. If the layer thicknesses are greater than the coherence length, then the layers are three-dimensional and behave as bulk material. In the opposite case the individual layers can be thought of as two-dimensional with respect to superconductivity. If in the multilayer the layers are weakly coupled then the system is thought of as quasi-two-dimensional. In this case the interlayer coherence length $\xi_{\perp}(0)$, which is perpendicular to the layers, is determined by the coupling between the layers. The relative size of $\xi_{\perp}(T)$ compared with the interlayer spacing Λ will determine the dimensional character of the system as a whole.

Table 6.1 also gives the superconducting properties that were examined for each system. This is usually at least the superconducting critical temperature as a function of the multilayer wavelength. For most of the systems the upper critical fields H_{c2} have been studied as well. H_{c2} is of particular interest because it acts as an indicator of the superconducting dimensionality of the multilayer. Qualitatively it is very well understood (see Ref.1 for a comprehensive discussion) and most current theoretical work in the field is concerned with quantitative calculations for specific systems. This is an area where experimentalists have had a chance for critical quantitative examination of the theory and vice versa.

Most researchers have examined other properties, besides superconducting, for their systems. It is of interest to correlate normal transport, magnetic, and other properties with the superconducting ones, either as a way of better understanding superconductivity in these samples or, on the other hand, in order to use superconductivity to learn about other effects. Localization effects in dirty systems, for example, have received considerable attention from the community and have been related to multilayer properties including superconducting ones.

We present a brief summary of some of the systems studied since the last review. Some of the newer work from Table 6.1 has not as yet been reported comprehensively in the literature and hence cannot be discussed in detail.

6.2.1 New Material Systems

Au/Ge

Au/Ge alternating ultrathin layered films were made by Akihama and Okamoto(5) and were shown to exhibit superconductivity. The multilayers were deposited by evaporation in an UHV environment. The substrates were cooled with liquid N_2 in order to prevent island formation, but the samples were subsequently brought to room temperature prior to measurement. It is apparent from the increase in resistivity with room temperature-annealing time that interdiffusion takes place to a great degree. However, the resistivity of the multilayers is still 100 times less than codeposited Au-Ge films.

The films consisted of 101 alternating Au(10Å)/Ge(13Å) layers. The onset of superconductivity occurs at 2.02K. The resistive superconducting transition is reported to have anomalous behavior, including reentrant behavior and electromotive force anomalies at the transition. It is not clear how much of this is due to sample inhomogeneities and current distributions in the measurements. The upper critical fields H_{c2} are anisotropic, but the temperature dependence is linear suggesting well-coupled layers and an overall 3D behavior.

Mo/Ni

The Mo/Ni system, which is of particular interest because of the possibility of magnetism in the Ni layers, has been studied extensively in the region of short modulation wavelength. Structural, magnetic and superconducting properties have been examined by Uher et al(7,9). These properties were studied as a function of layer thickness in the region 14Å<Λ<40Å. It is expected that in this region the properties of the multilayer would differ mostly from the bulk, because of competition between magnetic and superconducting behavior.

The samples were sputtered for equal thickness of Mo and Ni layers and for Mo layers three times that of Ni (in order to achieve thin Ni layers while coherent stacking of layers). For Λ>14Å layers are polycrystalline bcc Mo(110) and fcc Ni(111), while for shorter wavelengths the whole structure is amorphous, but still chemically ordered.

Structures with Ni layer thickness d_{Ni}<9Å show a superconducting transition whereas those with thicker Ni layers show no transition down to 15mK. Measurements of the magnetization **M** as a function of thickness show that **M** approaches zero at d_{Ni}=9Å, as well. This indicates that there is a loss of ferromagnetic order below 9Å Ni ($\sim$ 4 atomic layers). The T_c of the coherently layered films with thin Ni layers is observed to be $\sim$2K. The glassy structures show a $T_c \sim$0.5K.

The small wavelength multilayers (Λ=13.8Å and 16.6Å) exhibit large nonlinearities in the temperature dependence of the parallel upper critical field. Figure 6.2.1 shows the parallel and perpendicular critical fields and their ratios. The square-root-like temperature-dependence is suggested by the authors to be due to surface superconductivity effects in the anisotropic superconductor. Normally such behavior is associated with extreme quasi-two-

194

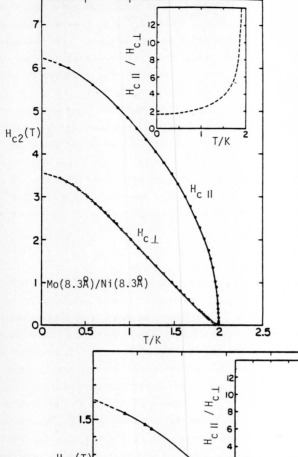

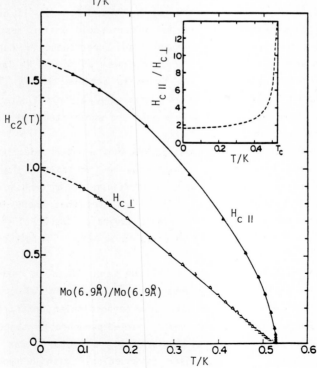

Fig. 6.2.1 Temperature
dependence of the
parallel $H_{c\parallel}$ and
perpendicular $H_{c\perp}$
critical fields for
Mo-Ni multilayers with
$\Lambda = 16.6\text{Å}$ (above) and
13.8Å (below). Insets
show critical field
anisotropy. X-ray
structural data
indicate that layering
is preserved in the
longer period
structure and show a
glassylike diffraction
pattern for the
shorter period sample.
[From Uher et al.(9)]

dimensionality of the system.

Mo/Sb

Mo/Sb multilayers were made by Asada and Ogawa(10). The multilayers, prepared by Knudsen cell and electron beam evaporation, are made up of textured polycrystalline bcc Mo and hcp Sb layers. X-ray diffraction shows only three low-angle peaks and no satellites for high-angle peaks, questioning the presumed structural order.

Superconducting transitions are observed at $\sim$5K. The T_c increases with decreasing wavelength, at constant ratio of layer thicknesses, up to a maximum of 5.95K. Bulk bcc Mo has a $T_c \sim 0.9K$, while amorphous Mo has $T_c \sim 8K$, and Sb is not normally superconducting, except under high pressure. It is not clear how the relatively high T_c of the multilayer can be obtained, since there is no indication that any amorphous interfacial region could be thicker than 10Å. The authors suggest that strained interfaces may be important for this effect.

Nb/RE (RE=Er, Tm, Lu)

Rare earth/Nb metallic multilayers have been grown by sputter deposition. Preliminary work on these Nb/Er, Tm, Lu systems has been reported by Greene et al(23). These systems with their varied magnetic structures of the rare earth (RE) metals are also good candidates for studying the interplay between long range magnetic and superconducting ordering.

The authors want to observe the proximity effect between magnetic and superconducting layers as a function of layer thicknesses. Nb forms no known compounds with the RE metals and is thus expected to have sharp interfaces with these. This has in fact been confirmed to be true, to within 10Å, by Rutherford backscattering experiments and low-angle X-ray diffraction. X-ray analysis also shows that for longer wavelengths ($\Lambda > 50Å$) the layers are textured polycrystalline bcc Nb and hcp RE crystals.

Magnetic ordering temperatures T_{FM} for Er and Tm are 18K and 32K, respectively, while Lu is not magnetic. There is a striking difference between the transition temperature T_c vs. Λ for multilayers with magnetic Er and nonmagnetic Lu, as shown in Fig. 6.2.2. In Nb/Lu T_c is depressed to 2.5K for Λ less than the Nb superconducting coherence length ($\xi_{Nb} \sim 500Å$) and then levels off. This is presumably due to disordered Nb layers for shorter wavelengths. In the Nb/Er system superconductivity is completely destroyed below $\Lambda \sim 200Å$. The authors also propose to study a RE-alloy/Nb multilayer system such that T_{FM} is below T_c, in order to possibly observe reetrant superconductivity.

196

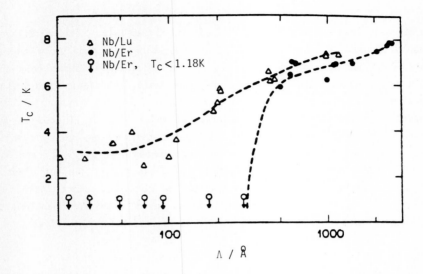

Fig. 6.2.2   Modulation wavelength Λ vs. superconducting
transition temperature T_c for Nb/Er (magnetic) and Nb/Lu (non
magnetic) multilayers grown on room temperature sapphire.   Note
the dramatic drop in T_c with decreasing Λ in the magnetic
system.   Open circles with arrows denote $T_c < 1.18K$, the lower
limit of temperature that we reach at present.   Dashed lines are
a guide to the eye.   [From Greene et al.(23)]

Nb/Ta

Broussard and Geballe have made Nb/Ta single-crystal multilayers in order
to establish how well such single crystals can be made and to measure their
superconducting critical fields(27).   These multilayers, prepared by sputtering
onto sapphire substrates at 800°C, were similar to the ones made previously by
Durbin et al(24,25) and Hertel et al(26).

Perpendicular critical fields $H_{c2\perp}$ exhibit positive curvature for all
samples, and this curvature increases with increasing Nb layer thickness.   This
appears to be explained by intrinsic curvature in H_{c2} of Nb (possibly due to the
anisotropic Fermi surface), which was also measured in the case of a thick pure
Nb film.

Dimensional crossover is observed in parallel critical fields for these
multilayers with appropriate wavelengths.   However, for samples with longer
wavelengths, i.e., d_{Nb}=290Å and 490Å, another transition in the critical field
behavior is observed at temperatures below the dimensional crossover, as shown

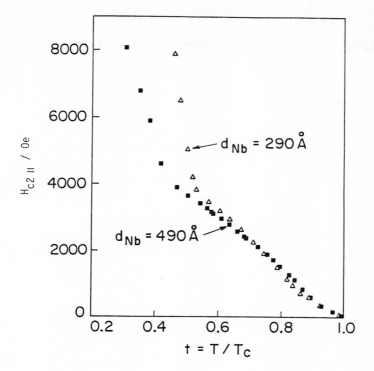

Fig. 6.2.3 Upper critical fields of Nb/Ta multilayers.
[From Broussard and Geballe(27)]

in Fig. 6.2.3. The angular dependence below this transition is very sharply peaked in the parallel direction (sharper than thin film or H_{c3} dependence). A possible explanation is offered by the work of Takahashi and Tachiki(60), discussed in Section 6.3, where the interplay between two superconducting layers plays an important role.

V/Ag

Multilayered V/Ag films have been studied by Kanoda et al(39-41). These films were prepared by UHV e-beam evaporation. They consist of textured bcc V and fcc Ag layers.

Multilayers were quasi-2D and 3D with respect to superconductivity. Upper critical fields were measured for a series of layer thicknesses and compared with current theories of Biagi, Kogan, and Clem(61), as well as Takahashi and Tachiki(62) with good quantitative agreement. This analysis and comparison of experiment and theory seem to be the most comprehensive to date.

Magnetic field penetration depth λ_{eff} as a function of layer thickness was also measured. It is observed that λ_{eff} first decreases with increasing multilayer period, but then turns upward. This is explained in terms of λ_{eff} for a single superconductor at short periods, while for longer periods the weak coupling of the V layers plays an important role. The competition between these two characters gives a minimum in λ_{eff}.

V/Fe

V/Fe multilayers are another system where the interplay between ferromagnetic and superconductivity was the object of study(42). Wong et al made these multilayers by evaporation in an UHV environment using a computer-controlled rotating table. To our knowledge comprehensive characterization of the multilayer order has not been reported in the literature.

These multilayers were made such that the Fe layer are only a few atomic planes thick. For small V layer thicknesses d_V the superconducting transition is rapidly quenched with increasing d_{Fe}. The authors point out that this T_c suppression is much faster in a homogeneous alloy, where only 5% Fe is required to destroy superconductivity, which indicates that the samples indeed have a composition gradient. For d_V about the size of the BCS coherence length for bulk V (440Å) T_c is suppressed to about 1K below the V bulk T_c=5.4K. A reported small increase in T_c at larger d_{Fe} does not seem to fit any theoretical framework.

Critical fields were measured and dimensional crossover was observed for samples with Fe layers of about 3 atomic planes. This is the first observation of dimensional crossover in an S/M system. 3D and 2D behavior are observed in the extremes of Fe layer thickness. In the 3D regime the critical fields are linear near T_c, but H_{c2} goes below H_{c2} at lower temperatures and this is unexplained. The authors suggest that magnetic scattering might be important for this.

V/Mo

Periodic and quasiperiodic (on same scales) Mo/V multilayers have been magnetron sputtered by Karkut et al(43-45). Karkut and coworkers have used a Fibonacci series as a generating rule for placing two different thicknesses of V layers in the multilayer. Superconducting properties were then examined to see if they were affected by the change in the order parameter due to a self-similar geometry of the layers.

The Mo/V layers were shown to be made up of textured polycrystalline bcc layers. High angle diffraction showed satellites due to either periodic or quasiperiodic layering.

For periodic multilayers the measured T_c values do not agree well with

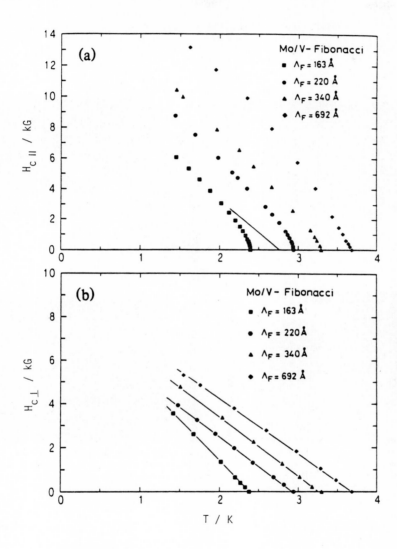

Fig. 6.2.4 Parallel and perpendicular upper critical fields of
four quasiperiodic Mo/V multilayers. The solid line in (a)
depicts the parallel upper critical field of a 70Å periodic Mo/V
multilayer. [From Karkut et al.(44)]

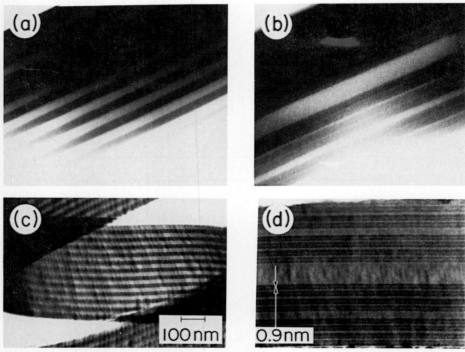

Fig. 6.2.5 TEM micrographs of cross sections of periodic (on the left) and fractal (on the right) MoGe/MoGe multilayers. [Courtesy of Robert Byers and Ann Marshall at Stanford University, from Ref. 47]

proximity effect theory calculations that assume an interfacial layer. Critical fields were also measured for these samples. The periodic sample with Λ=70Å shows linear T behavior near T_c. The critical fields for the quasiperiodic samples are shown in Fig.6.2.4. The parallel critical fields show some curvature even though the Mo layers there are only 15Å thick. This seems to indicate that the V layers are not coupled completely to give 3D-like behavior. From an experimental point of view it is not clear whether this observed critical field behavior is due solely to quasiperiodicity or just to the fact that there are 2 different V layer thicknesses in the multilayer (in the latter case this behavior would be seen even in a periodic arrangement of these).

MoGe/MoGe

Amorphous Mo-Ge alloys of two different compositions, one Mo-rich and the other Ge-rich, have been sputtered into multilayers by Matijasevic and Beasley(47). TEM cross-sections, shown in Fig.6.2.5, indicate that the interfaces are sharp and that the individual layers are continuous down to less than 10Å.

Samples were prepared in different layering geometries by varying the normal metal thickness (Ge-rich alloy) within the multilayer. Parallel upper critical fields have been measured and are shown in Fig.6.2.6. For a periodic multilayer the parallel field exhibits the expected linear behavior.

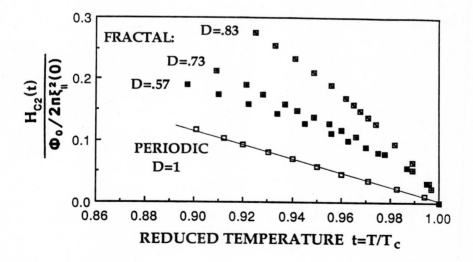

Fig. 6.2.6 Upper critical fields for periodic and fractal MoGe multilayers. [From Matijasevic and Beasley(47)]

However, when the normal metal thickness is varied and made thicker successively within the multilayer, such as the case for a fractal layering, then the critical field shows curvature. In fact it shows more curvature if the normal layer thickness increases by greater amounts across the multilayer for example, as the fractal dimension is reduced. The authors have considered geometries such that the nominal metal layers are: periodic, "doubly" periodic, random, and fractal. A Ginzburg–Landau–like scaling approach for the fractally layered samples seems to agree well with the critical field data.

In related work on MoGe multilayers, Missert and Beasley(48) have examined multilayers with varying compositions of the normal metal alloy. A fit of the T_c's for the multilayers to the proximity effect theory shows an increase in the superconducting layer transition temperature T_{cs} over the single film T_c. Also T_{cs} appears to vary with the composition of the normal metal layer. These results are as yet unexplained.

6.3 THEORETICAL CALCULATIONS

Although a general theory of superconducting multilayers does not exist, over the past decade and a half a coherent theoretical picture of the superconducting phase boundary for multilayers has emerged. This picture is based on the understanding of the proximity effect, tracing its roots back to the early work of de Gennes(63), and Josephson coupling of superconductors. Although the theoretical approaches have been on different levels, a standard model for understanding superconductivity in multilayered structures has been formed. Qualitatively, the model, which encompasses almost all the theoretical work on superconducting multilayers, uses the conventional understanding of superconductivity within the layers and matches the proper boundary conditions at the interfaces. We refer to a previous review(1) for a general description of this theoretical picture.

The development and refinement of this theoretical picture has afforded the research community an opportunity for direct comparisons between theory and experiment. In particular critical fields are of interest and have been used in this comparison. The critical temperature can be seen as the case when H=0. Critical fields are sensitive to the structure of the superconductor. For example, reduced dimensionality of the superconducting electrons as well as any stochastic element in the orbital motion of the electrons (such as boundary and impurity scattering) leads to an enhancement of the critical fields (for second order transitions).

There are several regimes that one can consider in doing a theoretical calculation for multilayers. One trivial case is when each layer is 3D in that the intrinsic coherence length ξ_s of the superconducting layers is much smaller than the individual layer thickness $\xi_s \ll d$. Here one recovers bulk behavior. A more interesting case is when the individual layers are quasi-two dimensional $\xi_s > d$. Now one must consider the coupling between the layers. One extreme is where the coherence length of the coupled system $\xi_\perp$ is short compared with the layer repeat distance Λ.   Then the superconducting layers are essentially decoupled and the system behaves two dimensionally. On the other extreme all the superconducting layers are strongly coupled($\Lambda \ll \xi_\perp$) and the system behaves as an anisotropic bulk, since it has uniaxial symmetry.   In between these two regimes one has quasi-2D behavior and crossovers between the regimes.   These regimes have been studied theoretically in order to understand the superconducting phase boundary and characterize it in the various regimes.

The newer theoretical work is also largely an extension of this standard picture and is of interest because it offers more direct quantitative comparison with the experiments. In order to place the discussion in a perspective and to understand the relative significance of these efforts, we present a brief historical sketch of the theoretical development before going on to new results. In Section 6.4 we will discuss some of the open theoretical questions.

6.3.1 Historical Development

The first regime to be analyzed was the coupled-layer, anisotropic 3D limit.   Dobrosavljevic(64), following the work of Tilley(65) on anisotropic superconductors, used a Ginzburg-Landau (GL) approach to an S/N or S/S' multilayer system.   She matched the superconducting order parameters as solutions of the GL equations for individual layers with the appropriate boundary conditions, and then reduced this to an anisotropic GL equation for the system as a whole.   By doing this she got the explicit expressions for $H_{c2\perp}$ and $H_{c2//}$.

The first approach to the 2D-3D crossover regime was offered by Lawrence and Doniach(66) and Kats(67) who modeled the superconducting multilayer as a stacked array of 2D superconductors coupled via the Josephson effect.   They justified the anisotropic GL model as well as derived a crossover in the superconducting fluctuations above T_c. As shown later by Klemm et al(68) this Josephson-coupled model predicts a divergence in the parallel critical fields when the temperature dependent effective perpendicular coherence length is on the order of the interlayer spacing $\xi(T) \sim \Lambda$.   This divergence, which is unphysical, is limited by interlayer pairbreaking effect in a magnetic field.

Klemm, Luther, and Beasley (KLB) have carried out a microscopic calculation of the critical field and included Pauli paramagnetic limiting as well as spin-orbit scattering(68).   They showed, for the first time in a realistic calculation, what the critical field would look like in a dimensional crossover regime. In this case superconductivity goes from bulk anisotropic behavior to 2D Pauli limited as the temperature is lowered.

Another approach to limiting the critical field was to include finite layer thickness, which was done by Deutscher and Entin-Wohlman(69). They included orbital pairbreaking for a thin film within a Josephson coupled multilayer, and they recovered a similar crossover in the parallel critical field.

In order to model their experiments, Ruggiero, Barbee, and Beasley (RBB) developed the de Gennes-Werthamer proximity effect theory to include the pairbreaking effect of the magnetic field for an SN system(21). This approach accounts for the differences in diffusivities and densities of states at the Fermi levels of the S and N layers. They calculated, in the dirty limit, the critical temperature in the presence of a field perpendicular to the layers near T_c and thus obtained an analytical form for the slope $(dH_{c2}/dT)_{T_c}$. Essentially the same formulation had been considered previously by Martinoli for an SN bilayer(70).

6.3.2 Newer work on critical fields

Takahashi and Tachiki (TT) have done calculations of critical fields(62) based on de Gennes' method(71). They considered a multilayer of two different layers and for each one they define a uniform density of states N, diffusion constant D, and electron-electron interaction constant V. TT examined the effects of each one of these parameters separately. They show that the difference of these quantities in the layers can give rise to a variety of temperature dependence of $H_{c2//}$ and $H_{c2\perp}$. Dimensional crossover in $H_{c2//}$ is reproduced, and TT give its systematic dependence on layer thickness, as well as the other parameters. $H_{c2\perp}$ is shown to have positive curvature which depends on these parameters.

Their work also makes some qualitatively new predictions about the temperature dependence of H_{c2}. They analyze their theory in the case when the multilayer is composed of two superconductors with different diffusion constants(60). Since the critical field is dominated by the component of the system with its highest value, they conjecture that the critical field will crossover from a solution for the superconducting order parameter centered in the high D material to the one with the low D as the temperature is lowered. Since this involves the vortices shifting from one layer into another, they predict a first order phase transition in the superconducting HT phase diagram as shown in Fig.6.3.1. This effect should be seen when the ratio of the diffusion constants exceeds a certain critical value.

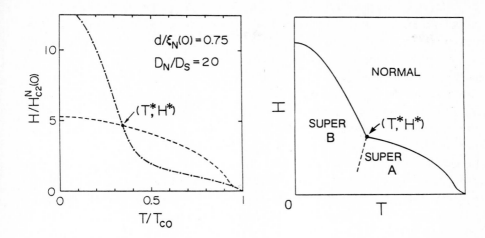

Fig. 6.3.1 Temperature dependence of the parallel critical
fields for a multilayer composed of two layers, N and S. The
dashed and dash-dotted curves on the left indicate the upper
critical fields when the order parameter nucleates in the N and
S layers, respectively. The observable critical field $H_{c2//}$ is
the higher one of the two critical fields at each temperature,
as shown on the right. The solid curves on the right indicate a
second order phase transition and the dashed curve indicates a
first order phase transition. The point (T^*, H^*) is a
multicritical point. [From Takahashi and Tachiki(60)]

TT have also calculated vortex pinning for the case of a multilayer with
layers of ferromagnetic and nonmagnetic superconductors(72). Based on a free
energy calculation they find that the critical pinning-force density is greatly
enhanced by pinning of the vortex lattice in the magnetic superconductor layers.

Earlier and independently Biagi, Kogan, and Clem (BKC) (61) used Usadel's
equations(73) to calculate H_{c2} . Their results are essentially equivalent to
the work of Takahashi and Tachiki. They have done a careful examination of the
positive curvature predicted by this theory and have used parameters that are
much easier to relate to experiment. This makes their work more attractive
comparisons. We note that this work justifies the model of RBB in a formal and
quantitative way.

Biagi et al have also examined their calculations for the case of a
m.lt la,er consisting of two superconductors, one of which has a higher critical
temperature but lower critical field(74). In this case the perpendicular
critical field will crossover from being dominated by high T_c material to low T_c
behavior (and higher critical field). This effect gives an upturn in H_{c2} at

lower T and is more pronounced as the layers are made thicker.

Kupriyanov has used the identical theoretical approach to calculate perpendicular critical fields for an SN bilayer(75). He analyzed the results in terms of two parameters, the suppression parameter γ characterizing the interface, and parameter $\gamma_M = \gamma(d_N/\xi_N)$ describing the characteristic distance over which the superconducting properties vary. For high values of γ and close to the critical temperature the degree of suppression of superconductivity due to the proximity effect is shown to be equivalent to suppression of superconductivity of a uniform film in a magnetic field.

Klemm has extended the KLB model for Josephson coupled superconducting layers to the case of SM multilayers(76). The superconducting layers are taken to be dirty with strong spin-orbit scattering. In addition he includes spin-flip scattering off the magnetic ions during interlayer tunneling. Critical fields are calculated for the cases of ferromagnetic, antiferromagnetic and paramagnetic ordering of the magnetic ions. Below the dimensional-crossover temperature a new type of spin ordering is predicted in parallel field, as well as cusps in the critical fields at the magnetic ordering temperatures.

For a long time some researchers have posed the question as to whether there is surface superconductivity in superconducting multilayers. Simonin has used a Ginzburg-Landau free energy variational approach to calculate transition temperatures as well as critical fields for multilayers(77). He assumed a functional form for the order parameter across the layers which is similar to Kittel's surface wavefunction(78) and then did a variational calculation. The assumptions in this model are somewhat different from the standard picture since the order parameter does not necessarily have translational symmetry in this model and in fact the order parameter is centered at the surface. The standard picture is reproduced within this model in the limiting case. However, it is not clear how valid are the assumptions for the functional form of the order parameter.

We also note here recent papers by Minenko(79) and Takanaka and Suzuki(80) on surface superconductivity in 3D anisotropic superconductors and thin films that may be helpful in understanding surface effects in superconducting multilayers.

6.4 CONCLUSIONS AND FUTURE DIRECTIONS

The number and varieties of artificially-structured superconductors that have been studied is now substantial. From the materials point of view, it is clear that a wide variety of superconducting multilayers and superlattices can be made. Fully-coherent single-crystal superlattice growth is rare, however.

Also, there appears to be a lower limit of about 10Å below which good layering of any kind cannot be achieved. As a consequence, it has not been possible in general to modify the normal state electronic properties of metals by layering. Some new and strained crystal structures in the multilayers with very thin layers have been observed (e.g., bcc Zr), but these new structures have not dramatically affected the properties of the multilayer. Similarly, evidence for compound formation at layer interfaces has been observed. It is often not well understood how important is the interfacial region itself on the superconducting properties.

Since layering has been readily achieved only on length scales above 10Å, the largest effects due to layering have been on those superconducting properties that depend on the superconducting pair wave function (or order parameter) and not the microscopic properties of the normal state. For example, it is very clear that layered superconductors have very different properties as type II superconductors than their bulk counterparts. The critical fields of superconductors are highly modified by layering as is now well demonstrated and well understood, at least for periodic layering. Measurements of H_{c2} are now perhaps most valuable as a materials diagnostic to establish the dimensionality and/or various material parameters. Some new features have been noted by Tachiki and Takahashi, however. The more recent work on nonperiodic layering (quasiperiodic and fractal layering) also opens up some new directions.

The role of H_{c3} and surface superconductivity (both with the vacuum and at internal boundaries) has not been properly elucidated. Some experimental anomalies have been attributed to surface superconductivity(9). A few theoretical works have addressed the question as well(77,79,80). Overall the situation is unclear, however, and surface effects remain an incompletely understood aspect of artificially-structured superconductors. Superconducting/magnetic multilayers also may hold some new possibilities. The interpretation of the results of Uher et al(9) on the H_{c2} of Mo/Ni multilayers are not resolved. The creation of artificially-structured magnetic superconductors analogous to the rare-earth borides may still be possible.

In contrast to H_{c2}, the behavior of artificially-structured superconductors below the phase boundary in the mixed state is mostly unexplored. Clearly there must be interesting flux pinning effects, commensurate-incommensurate transitions in the fluxoid lattice due to competing periodicity, and novel vortex dynamics in these materials. Also there is the question of the angular dependence of the mixed state in a layered superconductor. It is not clear why these issues have not received more attention. Suitable samples have been produced. The reason may be that the theory is much more difficult in this regime where the governing equations (e.g., the Ginzburg-Landau theory and its generalizations) become nonlinear. Experiments are still possible in any event.

Hopefully this region will be explored increasingly in the future.

Another relatively unexplored aspect of artificially-structured superconductors is the effect of layering on the microscopic properties of the superconductor. When the pair potential (which is related to the order parameter) is modulated, the single particle excitations of the superconductor also become modified. While the effects may not be as dramatic as those for the order parameter itself, they should be identified and explored in order to achieve a comprehensive understanding of the superconductivity of this class of materials.

Certainly one of the long term hopes of the researchers working on artificially-structured superconductors is that a breakthrough in high-T_c superconductivity or the identification of a new mechanism for superconductivity (e.g. excitonic mechanism) would be forthcoming. This has not yet happened. No dramatically high-T_c superconductors have been fabricated in this fashion nor have any compelling cases been proposed. Still, the structures being made continue to hold promise in this regard. At a minimum, it seems that this approach should at least provide model systems for testing some of the long-standing ideas regarding high-T_c superconductivity. The study of superconductors/semiconductor multilayers seems hardly exhausted. The control and material possibilities inherent in these techniques hopefully will provide some tests of the excitonic mechanism, even if they do not lead to higher T_c's.

The discovery of high temperature superconductivity in metallic oxides such as $(La-Sr)_2CuO_4$ (T_c=35K) and $YBa_2Cu_3O_7$ (T_c=90K) clearly changes the focus of the field of superconducting materials profoundly(81,82). Although superconductivity in these materials is not yet well understood, it appears likely that the layered nature of these materials plays an important role in their physical properties and possibly even in the existence of superconductivity itself. It may be that the understanding of superconductivity gained by the study of multilayered superconductors will be applicable to these new materials. More importantly, it may also be that the concept and approaches of artificially-structured superconductors can be applied at an atomic level to the synthesis of these materials. Only time will tell, but the prospects are tantalizing.

This chapter was prepared under the support of the U. S. National Science Foundation.

REFERENCES

(1) S. T. Ruggiero and M. R. Beasley, in "Synthetic Modulated Structures",
     L. L. Chang and B. C. Giessen eds., p. 365, Academic Press, New York, 1985.
(2) S. T. Ruggiero, Superlatt. Microst. 1 (1985) 441.
(3) I. K. Schuller and C. M. Falco, Thin Solid Films 90 (1982) 221.
(4) T. W. Haywood and D. G. Ast, Phys. Rev. B18 (1978) 2225.
(5) R. Akihama and Y. Okamoto, Solid State Commun. 53 (1985) 655.
(6) C. G. Granqvist and T. Claeson, Solid State Commun. 32 (1979) 531.
(7) C. Uher, R. Clarke, G. -G. Zheng and I. K. Schuller, Phys. Rev. B30 (1984)
     453.
(8) C. Uher, W. J. Watson, J. L. Cohn and I. K. Schuller, in "Layered
     Structures and Epitaxy", J. M. Gibson, G. C. Osbourne and R. M. Tromp eds.,
     North-Holland, Amsterdam, 1986.
(9) C. Uher, J. L. Cohn and I. K. Schuller, Phys. Rev. B34 (1986) 4906.
(10) Y. Asada and K. Ogawa, Solid State Commun. 60 (1986) 161.
(11) J. Geerk, M. Gurvitch, D. B. McWhan and J. M. Rowell, Physica 109, 110B
     (1982) 1775.
(12) D. B. McWhan, M. Gurvitch, J. M. Rowell and L. R. Walker, J. Appl. Phys. 54
     (1983) 3886.
(13) J. Guimpel, M. E. de la Cruz, F. de la Cruz, H. J. Fink, O. Laborde and
     J. C. Villegier, J. Low Temp. Phys. 63 (1986) 151.
(14) J. S. Chevrier, R. E. Somekh and J. E. Evetts, Mat. Res. Soc. 1986 Fall
     Meeting, p. 686.
(15) I. Banerjee, Q. S. Yang, C. M. Falco and I. K. Schuller, Solid State
     Commun. 41 (1982) 805.
(16) I. Banerjee, Q. S. Yang, C. M. Falco and I. K. Schuller, Phys. Rev. B28
     (1983) 5037.
(17) C. S. L. Chun, G. Zheng, J. L. Vincent and I. K. Schuller,   Phys. Rev.
     B29 (1984) 4915.
(18) Q. S. Yang, C. M. Falco and I. K. Schuller, Phys. Rev. B27 (1983) 3867.
(19) J. Guimpel, F. de la Cruz, J. Murduck and I. K. Schuller, Phys. Rev. B35
     (1987) 3655.
(20) S. T. Ruggiero, T. W. Barbee Jr. and M. R. Beasley, Phys. Rev. Lett. 45
     (1980) 1299.
(21) S. T. Ruggiero, T. W. Barbee Jr. and M. R. Beasley, Phys. Rev. B26 (1982)
     4894.
(22) M. G. Karkut, J. -M. Triscone and Ø. Fischer, Bull. Am. Soc. 32 (1987) 917.
(23) L. H. Greene, W. P. Lowe, W. L. Feldmann, B. Batlogg, D. B. McWhan and
     J. M. Rowell, Superlat. Microstr. 1 (1985) 545.
(24) S. M. Durbin, J. E. Cunningham, M. E. Mochel and C. P. Flynn, J. Phys. F11
     (1981) L223.
(25) S. M. Durbin, J. E. Cunningham and C. P. Flynn, J. Phys. F12 (1982) L75.
(26) G. Hertel, D. B. McWhan and J. M. Rowell, in "Superconductivity in d- and
     f-Band Metals, " p. 299, Kernforschungszentrum, Karlsruhe, Germany, 1982.
(27) P. R. Broussard and T. H. Geballe, Phys. Rev. B35 (1987) 1164.
(28) J. J. Lin, J. Cohn, F. Lamelas, H. He, R. Clarke, R. Merlin and C. Uher,
     Bull. Am Phy. Soc. 32 (1987) 916.
(29) J. Q. Zheng, J. B. Ketterson, C. M. Falco and I. K. Schuller, Physica 108B
     (1981) 945.
(30) Y. J. Qian, J. Q. Zheng, B. K. Sarma, H. Q. Yang, J. B. Ketterson and J. E.
     Hilliard, J. Low. Temp. Phys. 49 (1982) 279.
(31) W. P. Lowe and T. H. Geballe, Phys. Rev. B29 (1984) 4961.
(32) T. Claeson, J. B. Boyce, W. P. Lowe and T. H. Geballe, Phys. Rev. B29
     (1984) 4969.
(33) P. R. Broussard, D. Mael and T. H. Geballe, Phys. Rev. B30 (1984) 4055.
(34) H. Raffy, J. C. Renard and E. Guyon, Solid State Commun. 11 (1972) 1679.
(35) H. Raffy, E. Guyon and J. C. Renard, Solid State Commun. 14 (1974) 427,
     431.
(36) H. Raffy and E. Guyon, Physica 108B (1981) 947.

210

(37) T. Claeson, Thin Solid Films 66 (1980) 151.
(38) R. Clarke, F. Lamelas, C. Uher, C. P. Flynn and J. E. Cunningham, Phys. Rev. B34 (1986) 2022.
(39) K. Kanoda, H. Mazaki, T. Yamada, N. Hosoito and T. Shinjo, Phys. Rev. B33 (1986) 2052.
(40) K. Kanoda, H. Mazaki, N. Hosoito and T. Shinjo, Phys. Rev. B in press.
(41) K. Kanoda, H. Mazaki, T. Yamada, N. Hosoito and T. Shinjo, Phys. Rev. B35 (1987) 415.
(42) H. K. Wong, B. Y. Jin, H. Q. Yang, J. B. Ketterson and J. E. Hilliard, J. Low Temp. Phys. 63 (1986) 307.
(43) M. G. Karkut, D. Ariosa, J. —M. Triscone and Ø. Fischer, Phys. Rev. B32 (1985) 4800.
(44) M. G. Karkut, J. —M. Trinscone, D. Ariosa and Ø. Fischer, Phys. Rev. B34 (1986) 4390.
(45) J. —M. Triscone, D. Ariosa, M. G. Karkut and Ø. Fischer, Phys. Rev. B35 (1987) 3238.
(46) H. Homma, C. S. L. Chun, G. —G. Zheng and I. K. Schuller, Phys. Rev. B33 (1986) 3562.
(47) V. Matijasevic and M. R. Beasley, Phys. Rev. B35 (1987) 3175.
(48) N. Missert and R. Beasley, Bull. Am Phy. Soc. 32 (1987) 916.
(49) L. H. Greene, W. L. Feldmann and J. M. Rowell, Physica 135B (1985) 77.
(50) P. H. Schmidt, J. M. Vandenberg, R. Hamm and J. M. Rowell, in "Superconductivity in d— and f—Band Metals", H. Suhl and M. B. Maple eds., p. 57, Academic Press, New York, 1980.
(51) H. Yamamoto, M. Ikeda and M. Tanaka, Jap. J. App. Phys. 24 (1985) L314.
(52) J. Murduck, J. L. Vicent, I. K. Schuller and J. Ketterson, Bull. Am Phy. Soc. 32 (1987) 916.
(53) R. E. Howard, M. R. Beasley, T. H. King, R. H. Hammond, R. N. Norton, J. R. Salem and R. B. Zubeck, IEEE Trans. Magn. MAG—13 (1977) 138.
(54) B. Y. Jin, Y. H. Shen, H. Q. Yang, H. K. Wong, J. E. Hilliard, J. B. Ketterson and I. K. Schuller, J. Appl. Phys. 57 (1985) 2543.
(55) C. S. Baxter and W. M. Stobbs, Ultramicroscopy 16 (1985) 213.
(56) J. C. Bravman and R. Sinclair, J. Electr. Microsc. Tech. 1 (1984) 53.
(57) A. F. Marshall and D. C. Dobbertin, Ultratmicroscopy 19 (1986) 69.
(58) H. Q. Yang, B. Y. Jin, Y. H. Shen, H. K. Wong, J. E. Hilliard and J.B. Ketterson, Rev. Sci. Instrum. 56 (1985) 607.
(59) T. W. Barbee, Jr., in "Synthetic Modulated Structures", L. L. Chang and B. C. Giessen eds., p. 313, Academic Press, New York, 1985.
(60) S. Takahashi and M. Tachiki, Phys. Rev. B34 (1986) 3162.
(61) K. R. Biagi, V. G. Kogan and J. R. Clem, Phys. Rev. B32 (1985) 7165.
(62) S. Takahashi and M. Tachiki, Phys. Rev. B33 (1986) 4620.
(63) P. G. de Gennes, Rev. Modern Phys. 36 (1964) 225.
(64) Lj. Dobrosavljevic, Phys. Status Solidi B55 (1973) 773.
(65) D. R. Tilley, Proc. Phys. Soc. (London) 85 (1965) 1177.
(66) W. E. Lawrence and S. Doniach, Proc. Int. Conf. on Low Temp. Physics, 12th, E. Kanda ed., P. 361. Academic Press, Kyoto, Japan, 1970.
(67) E. I. Kats, Sov. Phys. —JETP 29 (1969) 897.
(68) R. A. Klemm, A. Luther and M. R. Beasley, Phys. Rev. B12 (1975) 877.
(69) G. Deutscher and O. Entin-Wohlman, Phys. Rev. B17 (1978) 1249.
(70) P. Martinoli and J. P. Meraldi, Solid State Commun. 9 (1971) 2123.
(71) P. G. de Gennes, Phys. Kondens. Mater. 3 (1964) 79.
(72) S. Takahashi and M. Tachiki, Phys. Rev. B35 (1987) 145.
(73) K. D. Usadel, Phys. Rev. Lett. 25 (1970) 507.
(74) K. R. Biagi, J. R. Clem and V. G. Kogan, Phys. Rev. B33 (1986) 3100.
(75) M. Yu. Kupriyanov, Fiz. Nizk. Temp. 11 (1985) 1244 [Sov. J. Low Temp. Phys. 11 (1985) 688].
(76) R. A. Klemm, Solid State Commun. 46 (1983) 705.
(77) J. Simonin, Phys. Rev. B33 (1986) 1700.
(78) P. G. de Gennes, Superconductivity of Metals and Alloys, p. 199, Benjamin, New York, 1966.

(79) E. V. Minenko, Fiz. Nizk. Temp. 9 (1983) 1036 [Sov. J. Low. Temp. Phys. 9 (1983) 535].
(80) K. Takanaka and M. Suzuki, J. Phys. Soc. Japan 55 (1986) 606.
(81) J. G. Bednorz and K. A. Müller, Z. Phys. B64 (1986) 189.
(82) M. K. Wu, J. R. Ashburn, C. J. Torng, P. H. Hor, R. L. Meng, L. Gao, Z. J. Huang, Y. Q. Wang and C. W. Chu, Phys. Rev. Lett. 58 (1987) 908.

Chapter 7

Theories on Metallic Superlattices

K. TERAKURA
University of Tokyo

7.1 INTRODUCTION

Theoretical and experimental studies are complementary in most fields of physics and this should apply particularly to the physics of artificial superlattices. The atomic configuration in a sample is more or less complicated, and theory alone cannot deal with such complexity satisfactorily. On the other hand, theory can propose some novel phenomena associated with an idealized structure, which is not necessarily accessible with present experimental techniques. Therefore such a theoretical prediction can serve as a guide for experiments. Of course, theoreticians are also making a lot of effort to analyze existing experimental data. Microscopic aspects of the observed pheonomena, which also may not be accessible by experiments, can sometimes be visualized by theoretical studies. In this context, the rapid progress of computers is very encouraging. One may regard a computer as a molecular-beam-epitaxy (MBE) machine, in which one can set up any sort of ideal superlattice and study its properties purely theoretically. Furthermore, one can even simulate the actual process of epitaxial growth. In fact, several detailed electronic structure calculations for rather complex systems and computer experiments by the Monte Carlo and molecular dynamics techniques have been performed and are also in progress with regard to the physics of artificial superlattices.

Experimental works so far may be categorized roughly as focusing on four areas, i.e., structural properties, magnetism, superconductivity, and elastic properties. Of these properties, the structural properties are the most basic, because the others are more or less affected by them. In this chapter, we review theoretical works on the magnetic, elastic and structural properties. (Superconductivity is discussed in Chapter 6.) The discussion focuses mainly on the material-science aspect of the metallic superlattices. Section 7-2 deals

with the interface magnetism paying particular attention to the alloying effect
at the interface and the associated local—environment effect. As examples, Fe/V
and Cu/Ni systems are discussed in some detail.   In Section 7.3, magnetism
enhancement at the interface and also at the surface are discussed.   The most
typical and extensively studied case is perhaps chromium.   Various other
possibilities have also been studied by heavy electronic—structure calculations.
Section 7.4 is devoted to discussions on the magnetic anisotropy, which is an
important subject not only from the standpoint of fundamental physics but also
application.   Elastic and structural properties are discussed in Section 7.5.
The supermodulus effect is certainly one of the most remarkable and novel
aspects of the metallic superlattice.   Some attempts to trace the origin of this
effect are reviewed.

7.2 ATOMIC CONFIGURATION AND MAGNETIC—MOMENT DISTRIBUTION AT INTERFACES

Before presenting experimental and theoretical studies for specific
systems, we discuss some general aspects of magnetic properties of
superlattices. Let us take z axis as the direction of film stacking. From the
many spin—density—functional(SDF) band calculations for metallic superlattices
and surfaces, we know that the electronic and magnetic structures of a given
layer are affected only by its first and second nearest—neighbor layers.   The
surface magnetism of Fe and Ni is a good example(1). Recently, the enhancement
in the magnetism at surface layers has been verified experimentally and
theoretically, as discussed in the next section, but the magnetic moment at the
layer next to the surface is very close to the bulk value. Therefore, if we
define a characteristic length ξ such that the electronic and magnetic
structures of a given layer are mostly determined by the neighboring layers
separated by less than ξ, it is only about the distance of the second—nearest
neighbor. Such a situation can be seen not only in layered materials but also
in many compounds and alloys(2,3).

Let us consider a superlattice whose concentration modulation along z
direction is given by c(z). We introduce another length scale $\lambda(z)$ by the
inverse of dc(z)/dz. If $\lambda(z) > \xi$, the electronic and magnetic properties of
the superlattice at z will be similar to those of a bulk alloy with
concentration c(z). Therefore, if the above condition is satisfied for all z,
the superlattice is nothing but a mixture of alloys with different
concentrations. (Of course, some properties, like transport, may reflect the
concentration variation over a much wider range.) Problems characteristic to
interfaces may occur, when the condition, $\lambda(z) < \xi$, is satisfied.

(i) <u>Fe/V</u>

Fe/V superlattice is certainly one of the systems studied most extensively both experimentally(4-9) and theoretically(3,10,11). Bulk Fe-V alloys show an interesting local-environment effect of magnetic moment and detailed theoretical studies on it are available(12-14). Attempts to derive information on the atomic configuration by analyzing the hyperfine-field (HFF) distribution have also been made for random alloys(15). Therefore, the interface magnetism of Fe/V superlattice is an interesting subject in view of the local-environment effect and the system may be a good candidate for studying the atomic configuration at the interface region. Detailed analyses, in fact, were performed by Hamada et al.(3,10). They carried out the SDF band calculations to find a relation between the iron magnetic moment and its local environment(3). The small magnetic moments at vanadium atoms are induced by the iron moments. Then the magnetic-moment distribution was related to the HFF distribution in a semi-empirical way. In principle, the analysis of this should also have been made with the first-principles calculation. However, this is an almost formidable task in practice, because of the insufficient accuracy in the SDF calculation with regard to HFF even for pure metals(16) and the complexity in the calculation for aperiodic systems.

Figure 7.2.1 shows the model superlattices of Fe/V, for which the SDF band calculations were carried out, and the calculated magnetic moment for an atom in each layer is summarized in Table I (3). In the actual superlattices, the (110) stacking is observed(5) and the (001)-stacking model in Fig.7.2.1 is simply to

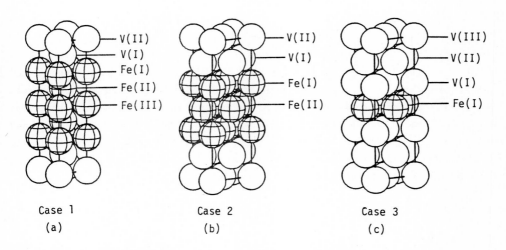

Case 1   Case 2   Case 3
(a)   (b)   (c)

Figure 7.2.1 Unit cells for model superlattices of Fe/V (from Ref.3). (001) stacking is assumed for Case 1 and (110) for Cases 2 and 3.

Table I. *Magnetic moment (in μ_B) in the muffin-tin sphere at each layer for the three cases shown in Fig.1 (from Ref.3).*

	(001) case 1	(110) case 2	(110) case 3
V(III)			0.01
V(II)	−0.07	−0.06	−0.01
V(I)	−0.24	−0.18	−0.12
Fe(I)	1.62	1.90	1.33
Fe(II)	2.39	2.20	
Fe(III)	2.21		

increase the samples to establish the relation between the magnetic moment and the atomic configuration. It is clear from Table I that the magnetic moment at a given iron atom decreases as the number of nearest−neighbor vanadium atoms increases.   The reduction of the iron magnetic moment at the interface was actually observed by neutron diffraction(4). It is important to note that the monolayer of iron can support a non−vanishing magnetic moment.   However, the significant reduction in the magnitude of the magnetic moment suggests that it is close to the collapsing of ferromagnetism. This is an important observation for analyzing the data by Hosoito et al.(4), who studied the case of a monolayer of iron in vanadium.   Within the limited number of samples studied, it was concluded that Hamada and Miwa's diagram(12) for the local−environment effect on the iron magnetic moment is applicable to the superlattices also, if the global alloy concentration is replaced with the concentration on the second−nearest neighbor shell of a given iron atom. This implies that the magnetic interaction range is limited only within the second−nearest neighbor distance as mentioned above.   For the magnetic moment of a vanadium atom, μ^V(in μ_B), the following simple equation can reproduce Hamada and Miwa's result(12) and also the result of SDF band calculation(3):

$$\mu^V = - 0.07(\bar{\mu}^{Fe} + \bar{\mu}^V) + 8.0 \times 10^{-4}\{(\bar{\mu}^{Fe})^3 + (\bar{\mu}^V)^3\}, \qquad (7.1)$$

where $\bar{\mu}^{Fe}$ and $\bar{\mu}^V$ are the total moments of Fe and V in μ_B on the first−nearest neighbor shell. Figure 7.2.2 shows the summary of the local−environment effect for the iron magnetic moments(3).   The subtlety of the monolayer iron case is now clearly seen.

In order to relate the magnetic-moment distribution to the HFF distribution, rather complicated empirical equations were used and the original paper(10) should be referred to for the details. Here we describe only the basic process of obtaining the microscopic information about the atomic configuration of the superlattice at the interface region. A trial concentration variation layer by layer was set up and the complete randomness in the atomic configuration in each layer was assumed. Then the HFF distribution was obtained for the particular atomic configuration. With trials and errors, the optimum concentration variation was obtained that could reproduce the overall features of the observed HFF distribution well. However, this process only was not sufficient to obtain a good agreement between theory and experiment. In order to proceed further, it was necessary to take account of the atomic-short-range order (ASRO). Hamada et al.(10) developed a new method for treating the ASRO even when concentration varies layer by layer. As the method may be of general use for superlattices, it is described briefly here.

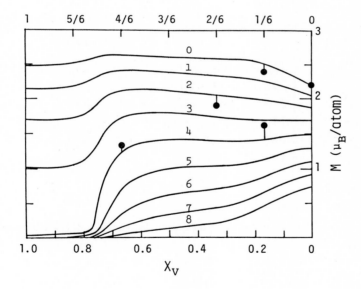

Figure 7.2.2 Atomic magnetic moments of Fe for random alloys (full curves) as functions of the V concentration, x_v. The numbers labelling the curves give the number of V atoms on the first-nearest-neighbor shell. As for the results for the layered materials (●), x_v is the local concentration of V in the second-nearest-neighbor shell.(from Ref.3)

Let us consider a superlattice composed of A and B atoms. We introduce a quantity $q_{ij}^{\gamma\nu}$ which is the probability of finding a γ atom at the i-th layer and a ν atom at the j-th layer simultaneously. Obviously the following equalities must be satisfied:

$$\sum_{\nu} q_{ij}^{\gamma\nu} = x_i^\gamma, \tag{7.2}$$

$$\sum_{\gamma} q_{ij}^{\gamma\nu} = x_j^\nu, \tag{7.3}$$

where γ and ν denote A or B and x_i^γ is the concentration of γ atom in the i-th layer. A new ASRO parameter β_{ij} is defined through the equation

$$q_{ij}^{\gamma\nu} = x_i^\gamma x_j^\nu + 4s^{\gamma\nu}\beta_{ij}x_i^A x_i^B x_j^A x_j^B, \tag{7.4}$$

with

$$s^{\gamma\nu} = \begin{cases} 1 & \text{for } \gamma = \nu \\ -1 & \text{for } \gamma \neq \nu. \end{cases} \tag{7.5}$$

It is easy to show that $q_{ij}^{\gamma\nu}$ expressed by Eq.7.4 satisfies Eqs.7.2 and 7.3. The second term of Eq.7.4 vanishes if either the i-th or j-th plane consists entirely of a single element. This is a reasonable result as expected. With this definition of the ASRO parameter, a positive (negative) β corresponds to a case of segregation (ordering). In the bulk alloy case, β is related to Cowley's parameter α by

$$\alpha = 4\beta \, x^A x^B. \tag{7.6}$$

Figures 7.2.3a and b show the final results for two different samples. It should be noted that the ASRO of a segregation case was concluded for both. This is contrary to the fact that the equilibrium Fe-V alloy has a tendency of ordering. Hamada et al. attributed the positive sign of ASRO in the superlattice to the actual deposition process. Let us imagine the situation when deposition of one species of atom has just been completed. In many cases the surface atoms are likely to form islands because of the bonding between them. In the next deposition process on this layer, the atomic configuration of the new substrate surface may not be disturbed so much, if the substrate temperature is kept low and the kinetic energy of the incoming particle is small. Then a positive ASRO is a natural consequence. We expect that this is the case of the Fe/V sample we examined. Actually, the substrate temperature

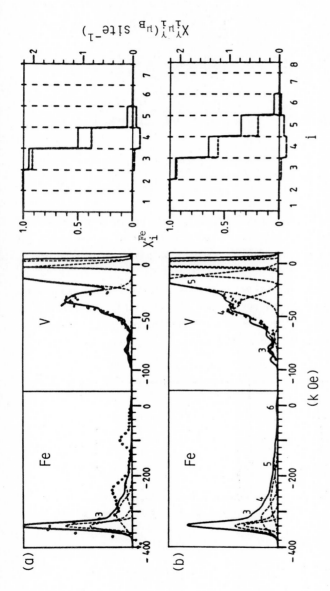

Figure 7.2.3 The HFF distributions for Fe and V, and concentration and magnetic moments of Fe and V of each layer for two different samples (from Ref.10). In the middle and left frames, full curves are the theoretical HFF distributions, each partitioned into the contribution from each layer, as indicated by broken curves. The numbers associated with the characteristic peaks of the broken curves denote the layers in the right frame. The histograms in the right frame show the atomic concentration and the magnetic-moment densities in each layer numbered by i along the horizontal axis: the full line is the Fe concentration (vertical axis on the left side), and the long and short broken lines are Fe and V moment densities, respectively (vertical axis on the right side). For the sample in (a), only the intra-layer ASRO of $\beta_{44} = 0.5$ is assumed, while for that in (b) $\beta_{44} = \beta_{55} = 0.3$ and $\beta_{45} = 0.4$ are assumed. The dots in the left and middle frames are the experimental data (Refs.4, 6 and 10).

(a)   (b)

Figure 7.2.4 Simulation patterns of atomic configurations for (a) β =0.0 (random alloy) and (b) β =0.5 (partial segregation) in a 50-50 alloy layer (from Ref.10).

was about 250 K, deposition rate was slow (0.2Å/sec), and the kinetic energy was also expected to be rather small in the process of synthesizing the samples.   In order to give a feeling of the effect of ASRO, simulation patterns are shown in Fig.7.2.4.

Another interesting case is an Fe monolayer in V metal.   An experiment corresponding to this case was carried out by Hosoito et al.(4).   According to the Mössbauer measurement, most of the iron atoms in the sample are nonmagnetic, while the SDF band calculation(3) predicts about 1.3 μ_B per iron atom.

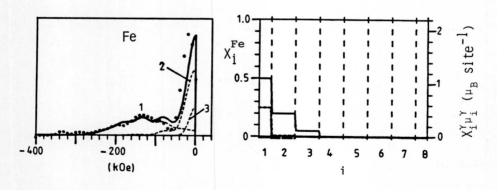

Figure 7.2.5 The HFF distribution for Fe, and concentration and magnetic moments of Fe and V of each layer for an Fe 'monolayer' in V metal (from Ref.10). The ASRO is assumed to be β_{11}=0.4 and β_{12}= 0.25. In the right frame, mirror symmetry is assumed with regard to the first layer. Other symbols are the same as in Fig.7.2.3.

This discrepancy can be resolved by assuming some distribution of iron atoms as indicated in Fig.7.2.5. The critical situation of an iron monolayer with regard to the magnetic stability is a crucial point in this analysis.

Very recently, a theoretical analysis for Fe/V superlattices was made by Elzain et al. based on the discrete-variational Xα method with small cluster models(11). Their results are qualitatively different from those of Hamada et al.(3,10). They predicted that the magnetic moment of Fe is almost independent of its local environment and furthermore, that it even increases slightly as the Fe atom is surrounded by more V atoms. The decrease in the HFF of Fe surrounded by many V atoms was attributed to a significant increase in the positive spin polarization in the s conduction electrons, which cancels the negative HFF caused by the d-electron spin polarization. Although this may be an interesting viewpoint, the fairly small size of the cluster (only 15 atoms altogether) makes one rather sceptical about the suitability of applying the model to actual superlattices. A fairly strong enhancement of the Fe moment even at the center of the cluster (2.85 μ_B in this cluster calculation versus bulk value of 2.15 μ_B) and a local-moment formation of V (with magnitude as large as or more than 3 μ_B in some cases) at the neighboring sites of the central Fe are certainly the size effect of the cluster model.

(ii) Cu/Ni

The first experimental study by Thaler et al.(17) concluded that the magnetism of Ni was appreciably enhanced in Cu/Ni superlattice. This stimulated extensive studies on this system experimentally and theoretically. The SDF band calculations have been done by Freeman's group to study the magnetic state of Ni at the interface of Cu/Ni(18-20). Figures 7.2.6a and b show some of the results. From Fig.7.2.6a, one can see that the magnetic moment of a Ni atom at the interface is reduced by about 30 % from the bulk value, but such a reduction is confined only to the first layer of the interface. A similar result was also obtained by Tersoff and Falicov(21). These calculations assumed a concentration modulation along the (111) or (100) direction. (The (111)-modulation was observed experimentally for superlattices.) It was also assumed that the composition of any given layer was purely Cu or Ni, although the experimental work was performed on samples with sinusoidal-composition modulation. Nevertheless, the theoretical result of reduced magnetism in Cu/Ni superlattice contradicted the experimental findings(17). It turned out that the interpretation of the experimental data was erroneous and later experiments supported the theoretical conclusions qualitatively. The result in Fig.7.2.6b for the (001) stacking can be compared with the recent experiment by Sill et al.(22), who measured the thickness dependence of the magnetic moment of Ni film sandwiched by Cu with a 500Å thickness. The theoretical result agrees well

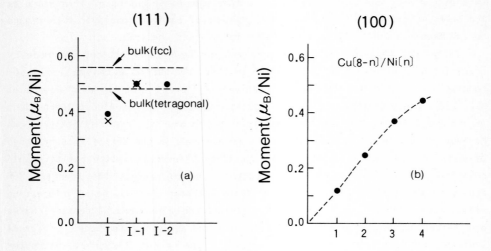

Figure 7.2.6 The calculated magnetic moments in Cu/Ni superlattices (data from Ref.19). (a) The layer-by-layer change in the magnetic moment of Ni in the (111) stackings of Cu[6]/Ni[6] (●) and Cu[3]/Ni[3] (×). I and I-i denote the interface and sub-interface. (b) The magnetic moment of Ni at the interface in the (100) stacking of Cu[8-n]/Ni[n]. Cu[i]/Ni[j] denotes a superlattice with i planes of Cu and j planes of Ni as a unit bilayer.

with the experiment. It is particularly interesting to note that both the theory and experiment indicate the presence of a fairly small magnetic moment for a monolayer film of Ni. However, one should be cautious about making a direct quantitative comparison between theory and experiment in this subtle case, because as we mentioned in the previous subsection, in the experiments the atomic configuration of films with monolayer thickness would not be ideally flat.

Although there was no disagreement in later experiments regarding the reduction in magnetism in Cu/Ni superlattice compared with pure Ni, there has been some confusion of the experimental data and their interpretation in the literature. For example, Flevaris et al.(23) claimed that the two-dimensionality was clearly evident from the linear-temperature dependence of magnetization for some samples with thin Ni-rich layers separated by relatively thick (6 or 8 layers) Cu and that for short modulation wavelength the moments of Ni-rich regions depend on the thickness of Cu-rich regions. However, these observations were criticized by Gyorgy et al.(24,25), who claimed that the magnetic properties could be interpreted simply by assuming that each atomic layer in the film behaves like a random alloy of the same composition. So far, theories have not been helpful enough to resolve the issue. A theoretical work such as CPA (coherent–potential approximation) calculation with a periodic concentration modulation is highly desirable. Also needed are experiments for samples with much sharper concentration modulations.

Although both Cu–Ni and Fe–V alloys are famous for their local–environment effect in magnetism, the two systems have qualitatively different aspects particularly near the critical concentration of ferromagnetism(13,14). In Fe–V alloy, the magnetic moment of Fe has a more or less localized character. Even in the non–magnetic concentration range, an Fe atom having more than four Fe atoms in its nearest neighbor shell is magnetic. Furthermore, the magnetic moment of an Fe atom is determined mostly by the local environment with the second nearest neighbor shell even near the critical concentration (see Fig.7.2.2). On the other hand, in Cu–Ni alloy, even if a Ni atom is surrounded by twelve Ni atoms in its nearest neighbor shell, it is non–magnetic (or at most has a vanishingly small magnetic moment) in the non–magnetic concentration range. In the ferromagnetic range near the critical concentration, the magnetic moment of Ni is supported by Ni atoms in a very much extended region. Therefore, in this particular case, the characteristic length ξ is very large. It would be very interesting to design a Cu/Ni superlattice so as to make use of this large ξ. One such possibility is to use $Cu_{1-x}Ni_x$ as one component and some magnetic element as another component of a superlattice. By adjusting x close to the critical concentration, the magnetic property of the sample may depend sensitively on x. A long–range exchange coupling between the magnetic layers may also be expected.

7.3 ENHANCED MAGNETISM AT SURFACES AND INTERFACES

The most remarkable example of enhanced magnetism at surfaces and interfaces observed so far is chromium. The bulk Cr is famous for its spin–density wave with a period of about 22 unit cells and a maximum magnetic moment of 0.59 μ_B(26). The Néel temperature is 312 K. Teraoka and Kanamori(27) developed a qualitative but rather general theory about the possibility of defect–associated enhancement of magnetism of Cr. For example, they pointed out the persistence of magnetic order at the surface even at 450 K. Allan studied this problem in more detail and predicted that the magnetic order in the (001) surface of Cr is ferromagnetic with a magnetic moment of 2.8 μ_B and that the magnetic ordering in the direction of surface normal is antiferromagnetic(28–30). The discussion was further extended by Grempel(31), who treated the surface plane separately from the rest and applied the functional integral method(32–36) to the finite temperature problem within the static approximation

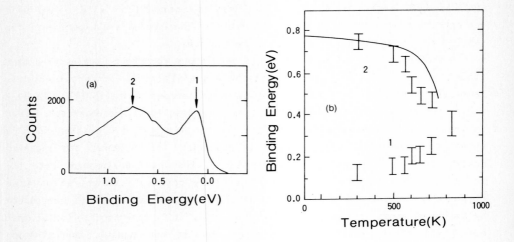

Figure 7.3.1 (a) The ARPES spectrum of normal emission from Cr(100) at 298 K and (b) the temperature dependence of the two peaks in (a) (from Ref.37). Bars are the experimental data and the solid curve is a theoretical result (from Ref.42).

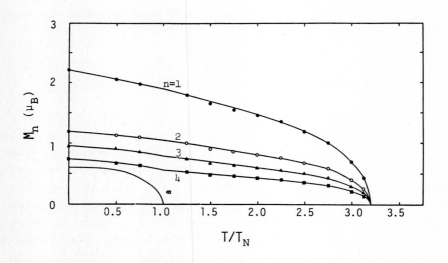

Figure 7.3.2 Temperature dependence of the average magnetization at the first four layers and in the bulk (from Ref.42). T_N is the bulk Néel temperature.

and the local-saddle point approximation to conclude that the ferromagnetic ordering at the surface would persist for temperatures up to about 900 K. These theoretical predictions were verified experimentally by Klebanoff et al.(37-40), who used the angle-resolved photoelectron spectroscopy (ARPES). Figure 7.3.1a shows the normal emission ARPES spectra (at 298 K) obtained by them and Fig.7.3.1b shows the temperature dependence of the binding energies of the two peaks in Fig.7.3.1a. These two peaks were concluded to be of surface-state origin because of their sensitivity to surface contamination. Detailed electronic structure calculation was carried out by Victora and Falicov(41) to analyze the experiment and peak 2 was assigned to nearly degenerate Γ_5-symmetry (+)-spin and (-)-spin surface states at the Γ point. (The label (+)-spin is used in reference to electrons whose spin magnetic moment is parallel to the magnetization of the surface layer.) From the temperature dependence of peak 2, T_{Cs}, the critical temperature of the surface magnetic ordering, was determined to be about 780 K. Very recently Hasegawa(42) analyzed the experiment carefully by extending Grempel's theory(31). The theoretical result for the temperature dependence of the state corresponding to peak 2 is shown in Fig.7.3.1b by a solid curve. The assignment of peak 1 has not been settled yet and we will not go into further details here. Figure 7.3.2 shows the temperature dependence of magnetization for the first four layers from the surface together with that for bulk(42).

Let us discuss the physical contents of the magnetism enhancement at the Cr surface. Cr crystallizes into the bcc structure. The d band, which is the basic ingredient in structural and magnetic properties, has a deep valley in the density of states near the middle of the band for the bcc structure. At the crystal surface, surface states and surface-resonance states are likely to appear in such a valley region. The important aspect of Cr is that the Fermi level is situated nearly at the bottom of the valley and therefore the density of states for the surface Cr atom becomes high enough to bring about a magnetic instability. The basic physics is similar to that for the surface reconstruction observed on the surfaces of Mo and W(43), which are in the same column of the periodic table as Cr. The 3d element has a larger Coulomb integral compared to the corresponding 4d and 5d elements due to the relatively well localized nature of 3d wave functions. This explains the reason why magnetic instability occurs in Cr but not in Mo and W. The enhanced magnetism of Cr is not confined only at the crystal surface but is also observed for alloys and microclusters. An addition of 20% of Al raises the Néel temperature to 900 K and the magnetic moment to 1.07 μ_B(44-46). The Curie temperature of microcluster of Cr also becomes as high as 800 K(47). All these phenomena are essentially ascribed to high densities of states at the Fermi level caused by defects in general(27). The basic concept here is the reduced coordination number for atoms situated near these defects. The localized states associated with the dangling bond will generally be located at the middle of the band and furthermore the reduced coordination will make the bandwidth narrow. Both of these effects enhance the density of states near the middle of the band.

The enhanced magnetism at the surfaces was also predicted for Fe(48) and Ni(49), and actually observed for Fe(50-52). According to the SDF band calculations, the magnetic moments of the surface layer of Fe(001) and Ni(001) are 2.98 μ_B(48) and 0.68 μ_B(49), respectively, which should be compared with the corresponding bulk values of 2.15 μ_B and 0.56 μ_B. The surface states do not play significant role in these cases, but the band narrowing caused by the reduced coordination number increases the density of states at the Fermi energy to enhance the magnetism. However, the fact that the magnetism enhancement is not so significant for Fe and Ni compared with Cr is reflected in the temperature dependences. In contrast to the Cr surface, the magnetization at the surfaces of Fe and Ni falls off more rapidly than the bulk magnetization as temperature increases(53).

It may be interesting to note that the magnetism enhancement of Fe can also be seen in alloys with non-transitional elements as in the case of Cr mentioned above. For example, the reduction in the magnetization per impurity atom is only 1.36 μ_B, and 0.97 μ_B for the cases of Ge and Sn, being less than the magnetic moment of an Fe atom(54). This implies an increase in the magnetic moment of Fe atoms at the neighboring sites of the impurity. Ni behaves in qualitatively different ways both at the interfaces and in alloys; the magnetization change per impurity atom of a non-transition element is about minus the impurity valency(55). Therefore, the magnetic moment at the surrounding Ni sites decreases. Similarly, enhanced magnetism has not been observed for Ni at the interfaces, though it was theoretically predicted at the free surface. The mechanism of the qualitative difference between Fe and Ni in the response to an impurity atom was explained by the author(56). Here we have emphasized some common aspects between the impurity problem and interface problem as pointed out by Tersoff and Falicov(21).

The possibilities of enhanced magnetism have been extensively explored through computer computations by Freeman's group(57,58). They have used the computer as a theoretician's MBE machine, setting up various atomic configurations and calculating (or measuring) their magnetic properties. This is probably one of the most promising directions taken in this field. From Table I of Ref.57 and that of Ref.58, we have selected only the results for Cr/Au cases and present them in Table II. It is clear from this table that a monolayer film of Cr, whether it is an overlayer on Au or sandwiched by Au or in a superlattice, the magnetic moment is enhanced by a factor of more than 5 compared with the bulk value. Furthermore, the interface Cr for a rather long-period superlattice has a magnetic moment of 1.65 μ_B, still being enhanced by a factor of about 3. It may be important to note the antiferromagnetic coupling between the neighboring Cr layers. There is also a possibility of antiferromagnetic ordering within each Cr layer(59). Some other cases studied so far are i) Fe overlayers on (001) surfaces of noble metals(57, 60,61), ii) monolayer film of Fe sandwiched by Ag and Au(57), iii) V overlayers on (001) surfaces of Ag and Au(57), and iv) monolayer film of V sandwiched by Ag(57). The subtle aspects of the V cases are interesting. Calculations showed that the (001) surface of V is nonmagnetic and the V of case iv) mentioned above is also nonmagnetic. However, the V of case iii) in the above has nearly 2 μ_B. In all these cases, a combination of a transition metal and noble metal was chosen, because, for this combination, the transition metal d band is separated from the noble metal d band energetically and therefore reduced coordination for the transition metals at the interface is virtually realized. Efforts to verify all of these predictions experimentally will be made in the future.

Table II. Theoretical layer-by-layer magnetic moments (in μ_B) of Cr (from Refs.57 and 58). S and (S-1) indicate surface and subsurface layers in the overlayer cases, while I and (I-n) indicate interface and sub-interface layers in the superlattice cases.

Magnetic moment of Cr

isolated monolayer	4.12
overlayers	
Cr[1] on Au(001)	3.70
Cr[2] on Au(001)	2.90 (S)
	−2.30 (S−1)
superlattices	
Cr[1]/Au[1] (antiferro)	3.00
Cr[1]/Au[3]	3.01
Cr[5]/Au[5]	1.65 (I)
	−0.79 (I−1)
	0.68 (I−2)

7.4 MAGNETIC ANISOTROPIES AT SURFACES AND INTERFACES

One of the striking properties of surface and interface magnetism is the anisotropy caused by the lowering of symmetry in atomic arrangement. Generally speaking, in-plane magnetization becomes favored as the thickness of a magnetic film decreases in order to reduce the magnetostatic energy. This is known as the shape anisotropy. However, several experimental observations have indicated that the magnetization becomes perpendicular to the plane in cases of ultra thin magnetic films, both for isolated films and for films sandwiched with non-magnetic materials. Some examples are Fe/Mg(62), Mn/Sb(63), Co/Pd(64), and Fe/Ag(65,66). Of course, there are some cases where the magnetization is more strongly confined within the plane, such as Pd/Fe(67) and Cu/Ni(25) (see also Ref.68 for other cases). Although some interesting observations have also been made with regard to the in-plane anisotropy(68), we will not discuss it because it is a higher order phenomenon compared with the out-of-plane anisotropy. (In order to avoid confusion, 'in-plane anisotropy' is defined as the anisotropy within the stacking plane and 'out-of-plane anisotropy' means the uniaxial anisotropy with regard to the plane normal; on the other hand, 'in- plane (out-of-plane) magnetization' means that the magnetization direction is within (perpendicular to) the stacking plane.) Perhaps, in the actual material, many complications, such as crystal imperfections and magnetic domains, may be involved. However, a fundamental understanding of the surface anisotropy even for ideally flat surfaces and interfaces is still lacking, particularly for metallic magnets. Only a few theoretical works are available. Before moving to the theoretical aspects, we will briefly describe some experimental data for Co/Pd(64) and Fe/Ag(65,66). Fe/Mg(62) and Mn/Sb(63) are discussed in Chapter 1.

Carcia et al.(64) studied the magnetization curves of superlattice of Co/Pd by applying external magnetic fields parallel and perpendicular to the film surfaces and found that the easy axis of magnetization switches from in-plane to out-of-plane as the thickness of Co film becomes less than 8Å. They estimated the total anisotropy energy K_u from the estimated areas between the parallel and perpendicular magnetization curves. K_u is defined as the difference between the parallel and perpendicular magnetization-energy densities and is expressed as

$$K_u = - (2 K_s + K_v d + 2\pi M_o^2 d)/\lambda, \tag{7.7}$$

where K_s is the interface anisotropy at each of the two interfaces in one bilayer period, K_v is the volume anisotropy of the Co layers, and $2\pi M_o^2$ is the shape anisotropy of the Co layers. d is the thickness of a Co layer and λ is the bilayer periodicity. Note that positive K_u means that the easy axis of magnetization is perpendicular to the film surface in the present convention.

λK_u vs d is shown in Fig.7.4.1. From this figure, K_s = -0.16 erg/cm^2 was obtained by extrapolating the curve to d = 0. Therefore, the surface anisotropy favors the out-of-plane magnetization. The figure also tells us that the transition between in-plane and out-of-plane magnetization occurs at about d = 8Å. Practically, the important aspect of the surface anisotropy of the Co/Pd superlattices is its stability against temperature. In fact, Carcia et al. found that heating of the sample with easy axis perpendicular to the film surface for two hours at 275°C in a vacuum did not change the magnetic properties, while heating at 400°C for two hours changes the easy axis to within the film surface. The heat treatment at higher temperatures may cause alloying at the interfaces and perhaps reduce the anisotropy in the atomic arrangement.

The argument about the surface magnetic anisotropy of Fe overlayer on Ag (001) is also interesting(65). As there is only 0.8 % mismatch between fcc Ag (001) and bcc Fe (001) surface nets after 45° rotation, an epitaxy of bcc Fe (001) plane is expected. The magnetic property of Fe overlayer was studied by spin- and angle-resolved photoemission. According to experiments, the spin averaged EDC (energy distribution curve) shows well defined two-peak structures indicating the exchange-split electronic structure for all cases studied.

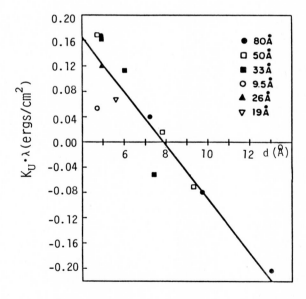

Figure 7.4.1 Total anisotropy energy K_u multiplied by the bilayer period λ versus the Co layer thickness d (from Ref.64). Symbols are for different Pd thicknesses and the solid line is a linear fit to the experimental data.

However, it was found that the in-plane spin polarization was negligibly small for films up to 2.5 monolayer (ML) coverage of Fe. For 5 ML film, the spin polarization was clearly observed. Two possibilities are conceivable concerning the absence of the in-plane spin polarization for thin overlayers. The first one is the absence of long-range order at room temperatures (at which the experiments were carried out) due to the reduced Curie temperature for thin magnetic films. The second one is the out-of-plane magnetization. In the case of Fe on Au(001), the Curie temperature for a comparable Fe coverage is higher than the room temperature(69). Therefore, the first possibility may not be applicable. The second one was supported by the detailed electronic-structure calculations by Gay and Richter(66), some details of which will be discussed later. In order to ascertain the result, it is desirable to measure the out-of-plane spin polarization experimentally.

The basic ingredients producing the surface-magnetic anisotropy are the spin-orbit interaction and the classical dipole-dipole interaction (magnetostatic energy). As mentioned above, the latter is the origin of the shape anisotropy and favors the in-plane magnetization. It is also important to note that the energy associated with the shape anisotropy is proportional to the thickness of the film so long as the film can be regarded to have an infinite two-dimensional extension compared with the film thickness. (Equation 7.7 already made use of this fact.) On the other hand, the anisotropy caused by the spin-orbit interaction is generally confined only within a few atomic layers at a surface or an interface. Therefore, even if this anisotropy tends to make magnetization perpendicular to the surface or interface, the in-plane magnetization will be eventually realized as the film becomes thicker. This argument implicitly assumes that the ferromagnetic exchange-coupling energy between magnetic moments is generally an order of magnitude larger than the anisotropy energy, so that variation of magnetization direction layer by layer is unlikely. (For the sake of simplicity, we will discuss only the 3d transition metals with ferromagnetic interaction.)

The first theoretical study on the surface-magnetic anisotropy was made by Néel for the localized moment systems(70). For the itinerant electron system, pioneering works were done by Bennett and Cooper(71) and Takayama et al.(72). Although the final result may not be reliable because of the lack of detailed information on the surface-electronic structure at that time, the work by Takayama et al. is instructive in several aspects. In the following, we present some formalistic aspects of treating the surface-magnetic anisotropy caused by the spin-orbit interaction. For the 3d transition metals, the spin-orbit interaction can be treated by a perturbation theory. A change in the one-electron band energy in the second order of the spin-orbit interaction is given by

$$\Delta E = - \frac{1}{2\pi} \int^{E_F} Im \ Tr\{GH'GH'\}dE, \qquad (7.8)$$

where G and E_F are the Green function and the Fermi energy of the unperturbed system(73) and H' is the spin-orbit interaction Hamiltonian expressed as

$$H' = A \sum_i \sum_{\sigma,\sigma'} \sum_{m,m'} <m\sigma|\vec{\ell}\cdot\vec{s}|m'\sigma'> a^+_{im\sigma} a_{im'\sigma'}. \qquad (7.9)$$

In H', A is the coupling constant as given by Eqs.2 and 3 of Ref.66 and about 0.08 eV for a Ni atom(72), for example. i denotes a site, m=1-5 corresponds to yz, zx, xy, x^2-y^2 and $3z^2-r^2$ orbitals respectively and σ is the spin index. $a(a^+)$ is an annihilation (creation) operator. $\hat{z}$ is the direction of the surface normal.

We define the anisotropy energy by

$$E_{anis} = K_s \cos\theta, \qquad (7.10)$$

with θ denoting the angle between the spin magnetization and the surface normal. Let us first consider the simplest case, where the spin-orbit interaction is taken into account only at a particular site i. Then a manipulation similar to that of Takayama et al.(72) leads us to

$$K_s = (A^2/4)\{F_i(2,1) + 4F_i(3,4) - F_i(1,3) - F_i(1,4) - 3F_i(1,5)\}, \qquad (7.11)$$

where

$$F_i(m,m') = \sum_\sigma \{F_i(m, \sigma; m', \sigma) - F_i(m, \sigma; m', -\sigma)\}, \qquad (7.12)$$

$$F_i(m, \sigma; m', \sigma') = - \frac{1}{\pi} \int^{E_F} Im\{G^\sigma_{im}(E)G^{\sigma'}_{im'}(E)\}dE. \qquad (7.13)$$

The Green function G^σ_{im} is related to the local density of states $n^\sigma_{im}(E)$ associated with the site i, orbital m and spin state σ by the following equation:

$$n^\sigma_{im}(E) = - \frac{1}{\pi} Im \ G^\sigma_{im}(E). \qquad (7.14)$$

If the local symmetry at the site i is cubic, the following equalities

$$F_i(2,1) = F_i(3,1), \qquad (7.15)$$

$$F_i(1,4) = F_i(3,4) = F_i(1,5), \qquad (7.16)$$

hold and K_s vanishes. As mentioned in Section 7.2, the effect of a surface or an interface on the local electronic structure extends only up to one or two layers. Therefore, the surface-magnetic anisotropy is confined also within the same narrow region. Even for this simplest case, it is hard to say more about the general aspect of K_s without doing the actual calculations.

In a real situation, where the spin-orbit interaction must be taken into account at all magnetic sites, the anisotropy energy is given by two different contributions. The first one is simply the sum of Eq.7.11 over the magnetic sites. Another contribution comes from the inter-atomic interaction, which is analogous to the non-local spin susceptibility(74) but is more complicated because of the matrix element of spin-orbit interaction. Takayama et al.(72) adopted the k-space analysis rather than the real-space one that we have adopted here. The real-space approach described here may be more suitable to obtain a physically transparent picture of this problem.

Very recently, Gay and Richter(66) carried out calculations of surface-magnetic anisotropy energy for monolayer films of V, Fe, and Ni based on the SDF-band calculations. They set the lattice constant for all three systems equal to that of Ag, in order to simulate the epitaxial overlayers on Ag. They adopted a perturbational approach for the degenerate case. The anisotropy energy was expressed as

$$E_{anis} = E^{(0)} + E_z^{(2)}\alpha_z^2 + E_{xz}^{(4)}(\alpha_x^2\alpha_z^2 + \alpha_y^2\alpha_z^2) + E_{xy}^{(4)}\alpha_x^2\alpha_y^2, \qquad (7.17)$$

where α_x, α_y, and α_z are direction cosines and $\hat{z}$ is perpendicular to the film plane. The four parameters were determined by calculating E_{anis} directly by taking four different directions of spin magnetization. The fourth order terms are an order of magnitude smaller than the second order one and we will present here only the results for the latter. The calculated values of $E_z^{(2)}$ ($= K_s$) are -0.38, 4.00, and -0.06 in meV/atom for Fe, Ni, and V respectively. The magnetic moments are 3.20 (Fe), 1.04 (Ni), and 3.00 (V) in μ_B. As the shape anisotropy is about 0.3 meV/atom both for Fe and V, Fe is the only case where the magnetization is aligned perpendicularly to the film surface. As mentioned above, the result was used to explain the experiment by Jonker et al.(65). The value of K_s for Fe in erg/cm^2 is -0.73, which is 4.6 times larger than that for Co/Pd(64) mentioned above.

7.5 ELASTIC AND STRUCTURAL PROPERTIES

(i) Elastic anomalies

In some metallic superlattices (Au/Ni(75), Cu/Pd(75), Cu/Ni(76), and Ag/Pd(77)), the biaxial elastic modulus Y[111] is significantly enhanced (by 200–400%) when the modulation wavelength becomes short (about 20Å). This is sometimes called the supermodulus effect. In contrast to the above cases, Cu/Au and Ag/Au do not show the supermodulus effect and furthermore there are some examples which exhibit a softening rather than hardening in the elastic moduli. Nb/Cu(78), Mo/Ni(79) and V/Ni(80) belong to the latter group. Table I of Ref.81 is a good summary of a variety of experimental data. As no decisive theoretical interpretation is available to the microscopic origin of these elastic anomalies, we will simply try to give an overview of the present status in this field.

Let us first summarize some characteristic aspects of the supermodulus effect: i) The supermodulus effect is sensitive to the wavelength λ of composition modulation and shows a peaking around λ of 20Å (76,77) (see Fig.7.5.1a for Cu/Ni); ii) The enhancement appears to be proportional to the square of the amplitude of the composition modulation (75–77)(see Fig.7.5.1b for Cu/Ni); iii) The stress–strain relation does not obey Hook's law but a monotonic decreasing function of the strain (75–77) (see Fig.7.5.1c for Cu/Ni); iv) There is a strong correlation between the λ–dependence of the elastic modulus and the magnitude of the variation in planar spacing along the modulation direction(77); v) The specimens having a very poor (111) texture do not show the supermodulus effect; vi) For the (001) epitaxial layers of Cu/Ni foils, the enhancement of Y[001] is not so drastic as that of Y[111] (see the arguments below).

Some theoretical attempts have been made to elucidate the microscopic origin of the supermodulus effect. One of the possibilities proposed so far is the Fermi-surface nesting (FSN). By using Fermi surfaces of the average composition alloys in the Cu/Ni and Ag/Pd systems, Pickett(82) showed that the superlattice Brillouin zone contacts with the Fermi surface for the particular λ at which the supermodulus effect is observed. Furthermore, he made a guess of the Fermi surface of Cu/Au alloy of 50-50 composition, and showed that the FSN would not occur in Cu/Au for any modulation wavelength. The same viewpoint was taken by Wu(83), who tried to calculate the change in the screened ion–ion interaction by the FSN effect. There are some observations which may support this mechanism. First, Tsakalakos and Hilliard(76) pointed out that the critical λ of 17Å for Cu/Ni is close to the wavelength predicted from the positron annihilation measurements for a concentration wave to be induced due to the FSN. Second, the measured thermo-emf (electro motive force) showed that the emf, which is closely related to the electron density of states at the   Fermi

234

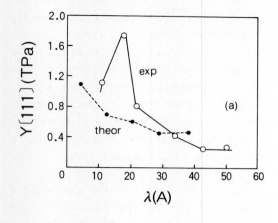

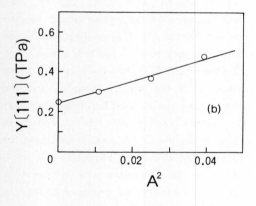

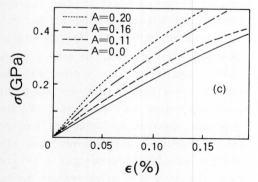

Figure 7.5.1 Biaxial elastic modulus Y[111] of Cu/Ni super-lattices (from Ref.76). (a) Y[111] vs. modulation wavelength for a 50 at.% modulation amplitude. The average composition of the samples is 50 at.% Cu. Theoretical result from Ref.88 is also shown.

(b) The variation of Y[111] as a function of the square of the modulation amplitude. The average composition of the samples is 50 at.% Cu and the modulation wavelength is 17.2 Å.

(c) Stress vs. strain curves of four samples with different amplitudes of composition modulation. The average composition of the samples is 45 at.% Cu and the modulation wavelength is 17.2 Å.

energy, also reached a maximum near the critical λ value both in Cu/Ni and Ag/Pd(84,85). Although the discussions are very interesting, more quantitative theoretical treatments are required to check the validity of the FSN mechanism. Unfortunately, however, the first-principles calculation of elastic moduli is rather difficult in general, even for pure metals. Before closing the arguments on the FSN mechanism, it may be worthwhile to cite the fundamental question addressed by Clapp(86): The FSN is generally regarded as the stabilization mechanism of a particular atomic configuration, but then "why are the modulated structures with the critical λ not stable with respect to the homogeneous alloy?"

An alternative view is to regard the coherent strain as the origin of the elastic anomalies. Tsakalakos and Jankowski(87) studied theoretically the effect of tetragonal strain on the elastic properties of Cu. First of all, they claimed that the supermodulus effect had been observed also in Y[001] for the (001) texture of Cu/Ni(76); a composition modulated Cu/Ni foil of 66% Cu has Y[001] of 0.23 TPa (1TPa = 10^{13}dyn/cm^2), which is about 50% greater than that of a bulk Cu/Ni alloy of the same Cu composition. (Note, however, that in Ref.76 the magnitude of enhancement was estimated to be about 20% and that it was attributed to the presence of (111) grains in the (001) texture.) They showed that C_{11}, C_{12}, and C_{66} are sensitive to the tetragonal strain and that Y[001] could be enhanced by 100% for 3% compressive strain. (In their convention, compressive strain is the strain that should exist in the Cu layers in Cu/Ni systems.) As for the decrease of the elastic modulus of ultrathin layers, which was sometimes taken as an important aspect favoring the FSN mechanism, they referred to the interdiffusion between layers which would reduce the amplitude of the concentration modulation. In an ordinary metallic material, any sort of mechanical deformation can change the elastic moduli by at most 1 or 2%, because the maximum attainable elastic strain is usually less than 1%. However, in metallic superlattices, a much larger elastic deformation may be possible because of the thin layer thickness. The importance of strain in the supermodulus effect was demonstrated also by Imafuku et al.(88), though they treated incoherent structures rather than coherent ones. They first constructed models of atomic arrangements in Au/Ni and Cu/Ni superlattices with five different periodicities (1-layer/1-layer, 3/3, 5/5, 7/7, and 9/9) as follows : The fcc (111) planes of each constituent metal were stacked such that the lattice spacings were the same as in the corresponding bulk crystals. Then the structure was relaxed by means of the molecular dynamics; in the relaxation process, the interatomic potential was assumed to be of the Morse type. The atomic arrangements obtained this way are necessarily incoherent. Interestingly, they found that although the intra-plane relaxation is very small except at the interface, the inter-plane relaxation is significantly large even in the region away from the interface. This aspect may be related to the experimental observation of the correlation between the supermodulus effect and the variation in the planar spacing as mentioned earlier. The biaxial modulus calculated with the model atomic arrangements does show an enhancement for superlattices with shorter periodicities, which is qualitatively consistent with the experiments(75,76). They also found that the initial atomic arrangements before the molecular-dynamics relaxation did not show any enhancement of elastic moduli. However, because the agreement between calculated and observed biaxial moduli is not satisfactorily good in a quantitative sense, it is not quite clear whether the essential ingredients of the supermodulus effect were correctly included in the theory. In fact, some problems may be conceivable. Among them, the most serious one may be the assumption of the incoherent structures even for cases with very thin layer thickness.

As was mentioned already, Nb/Cu(78), Mo/Ni(79), and V/Ni(80) show a softening in the elastic moduli in short period cases in contrast to the hardening in the supermodulus effect. First of all, one should note that these superlattices are composed of metals with bcc and fcc structures in the bulk crystals. Actually, x-ray measurements confirmed stackings of fcc (111) and bcc (110) in the layers of fcc and bcc metals, respectively, for rather long period cases (for example, $\lambda > 16\text{Å}$ in the case of Mo/Ni with equal layer thickness). In such cases, one may expect that instability in this stacking of layers may occur when the periodicity decreases. As the basic physics is common to all three cases and furthermore Mo/Ni has been most extensively studied, we discuss only the Mo/Ni case. Figure 7.5.2a shows the reciprocal of the average lattice spacing versus the periodicity λ of the Mo/Ni superlattices(79). One can see that the expansion of the lattice spacing occurs for small λ and that the effect is more pronounced for samples with a larger content of Ni. This implies that fcc (111) stacking of Ni comes close to an instability. The results for the equal-layered samples indicate that $\lambda = 16.6\text{Å}$ is the critical wavelength; the average inter-layer spacing expands about 2% and then suddenly contracts as λ comes down and crosses this critical value. (Note, however, that the in-plane strain is fairly small at the critical wavelength, 0.6% expansion in the Ni spacing and 0.5% contraction in the Mo spacing.) Figure 7.5.2b shows the surface-wave velocity versus the periodicity λ for three cases corresponding to those in Fig.7.5.2a. It is clear from Fig.7.5.2 that there is a close correlation between the inter-layer spacing and the surface-wave velocity and that the softening in the surface-wave velocity originates mostly from the strain in the Ni layers. This was further proved by a molecular-dynamics calculation of phonon velocity for strained thin Ni films(89). Not only the variation of phonon velocity as a function of the interplanar distance but also the critical interplanar distance with regard to the lattice instability were

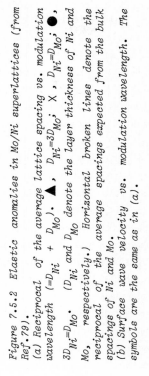

(b)

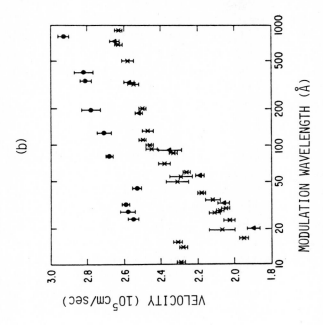

(a)

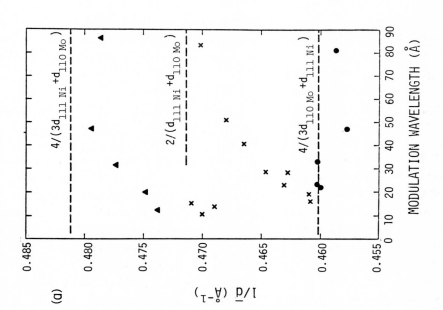

Figure 7.5.2 Elastic anomalies in Mo/Ni superlattices (from Ref.79).
(a) Reciprocal of the average lattice spacing vs. modulation wavelength (=D_{Ni} + D_{Mo}). ▲, D_{Ni}=3D_{Mo}; X , D_{Ni}=D_{Mo}; ●, 3D_{Ni}=D_{Mo}. (D_{Ni} and D_{Mo} denote the layer thickness of Ni and Mo, respectively.) Horizontal broken lines denote the reciprocal of the average spacings expected from the bulk spacings of Ni and Mo.
(b) Surface wave velocity vs. modulation wavelength. The symbols are the same as in (a).

correctly reproduced.

In both cases of hardening and softening in the elastic moduli, there are some theoretical and experimental arguments that the strain, particularly the inter-plane strain, is the origin of the elastic anomalies. However, the origin of the strain is not well understood. By using simple interatomic potentials, Imafuku et al.(88) obtained a fairly large inter-plane strain, which is never related to the FSN mechanism. Nevertheless, because a quantitative agreement of their results with experiments is lacking, we cannot totally exclude the possibility that the FSN mechanism plays a role.

(ii) Structural problems

In order to have deep understanding of properties of superlattice in the atomistic level, information of the atomic configuration is substantial. Although neutron and x-ray diffraction experiments are powerful techniques to obtain information about periodicity, it is rather difficult to identify local-atomic configuration by using the experimental techniques presently available. We presented in Section 7.2 an example of how such information can be obtained by analyzing hyperfine-field distributions through NMR and Mössbauer measurements. But this is a rather indirect method for this purpose. Considering this situation, we again point out the great potential ability of computer simulations for playing a significant role. We already mentioned two examples of molecular-dynamics simulations(88,89) for elastic properties of superlattices. In this subsection, we present some additional examples.

The basic structural problems that we will briefly discuss here concern the layer stacking between two constituents i) which have the same crystal structure but different lattice parameters in their bulk states, ii) which have different crystal structures (for example, fcc and bcc), and iii) which have different but similar crystal structures (for example, fcc and hcp). Finally a computer simulation of an actual MBE process is described.

As for the first problem, we mentioned earlier the work by Imafuku et al.(88). Dodson(90) also studied the stability problem of two-dimensional slabs composed of two different lattice systems with different lattice parameters by using a Monte Carlo technique. In the model, each lattice system consisted of 12 layers. One layer has 120 atoms and the atoms interact with their nearest neighbors with the Lennard-Jones (LJ) or modified directional LJ potential. His arguments started from a completely opposite situation to that of Imafuku et al.(88). The initial coherent structure was set up by compressing or expanding the two constituent systems isotropically so that an atomically perfect interface was formed. Then the whole system was disposed to a free boundary condition and the initial state was relaxed through the Monte Carlo technique for 0 K. If the coherent strained-layer state is unstable, the system would end up with a defective interface and lattice. However, the analysis showed that any sign of loss of registry did not appear for both the ordinary LJ and the directional LJ potentials up to 15% lattice mismatch, the largest mismatch value considered. Although the system-size dependence was not studied, the result suggests that once a coherent structure is formed, it is fairly stable.

Ramirez et al.(91) studied the bcc(110)-fcc(111) epitaxy by using a simple rigid-lattice model. In the model, the substrate consisted of three bcc (110) planes stacked properly as part of the bulk bcc structure and one fcc (111) plane. The bcc plane was infinite, while the fcc plane was assumed to be a disk of finite radius. Introducing two types of potentials, the LJ potential and an exponentially decaying potential, they calculated constant-energy contours for the stacking of the bcc and fcc layers in the P-θ plane, where P is the ratio of the fcc/bcc lattice parameters and θ the angle between the fcc (111) principal axis and that of bcc (110). They found two prominent minima for both types of potentials, one at $\theta = 5°$ and P=1.33 and the other at $\theta = 0$ and P=1.16. The calculated results and a comparison with some experimental results are shown in Fig.7.5.3. Despite the simplicity of the model, there is a reasonably good correspondence between the theoretical prediction and the experiments for

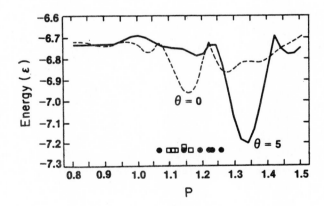

Figure 7.5.3 Energy of stacking vs. ratio of fcc/bcc lattice parameters for the Lennard-Jones potential (from Ref.91). P=fcc/bcc lattice parameter.   The symbols along the x-axis indicate specific systems.   From left to right, ●: Nb/Ni, □:Nb/Cu, □:W/Ni, □:Mo/Ni, ●:W/Cu, □:Mo/Cu, □:V/Ni, ●:V/Cu, ●:Cr/Ni, ●:Cr/Cu and ●:Fe/Cu.   The systems denoted by □ exhibit superlattice growth, while those by ● do not.

the curve of θ = 0.   As for the curve of θ = 5°, no experimental data is available which exhibits superlattice growth around P=1.33.   One may guess that a possibly random stacking around θ = ±5° might make layered growth difficult.

The third problem is related to the phenomena of polytypism(92). Cunningham and Flynn(93) studied the layer stacking in Ru/Ir superlattices. In their bulk crystals, Ru is hcp and Ir is fcc with interatomic distances of 2.706Å and 2.715Å, respectively.   As the close-packed hexagonal planes of the two lattices have closely similar spacings, a high quality of epitaxial growth is expected and furthermore, it is interesting to see how the layer stacking in each constituent metal is influenced by the contact with the other.   They prepared three samples: (a) one periodicity consists of 11 atomic planes of Ru and 5 of Ir (denoted as Ru[11]/Ir[5]), (b) Ru[11]/Ir[10], and (c) Ru[10]/Ir[15]. X-ray diffraction analysis showed that Ir is entirely hcp for sample (a), about 3 layers of Ir may be hcp for sample (b), and Ir is fcc for sample (c).   The energy cost to convert fcc to hcp is fairly small and a small strain may be sufficient to transform Ir to hcp.   In fact, they found that the superlattice sidebands exhibited a strong asymmetry which was associated with strain in the stacking direction.   There is also a possibility that the charge transfer between Ru and Ir is important.   A detailed electronic structure calculation would be very helpful in this context, although the energy involved here is fairly small.   Redfield and Zangwill(94) presented a general consideration on this subject with a simple model.

The final topic covered here is a computer simulation of the MBE process. Although the works by Imafuku et al.(88) and Dodson(90) may elucidate some fundamental aspects of the atomic configuration in superlattices, their approach may not be suitable to simulate the actual layer-by-layer crystal growth.   The atomic configuration depends not only on the static properties but also the dynamical properties, because the system may not be in a thermal equilibrium. We mentioned in Section 7.2 that the atomic configuration at the interface of Fe/V superlattice would strongly depend on the deposition process.   We believe that the molecular-dynamics simulation is a very promising approach for elucidating the detailed aspects of the MBE process.   One example of such a study is the work by Schneider et al.(95), though it dealt with the epitaxial growth of a single element rather than the superlattice growth by two different elements.   They assumed the LJ type inter-atomic potential between particles

$$V(r) = 4\varepsilon\{(\sigma/r)^{12} - (\sigma/r)^6\}. \tag{7.18}$$

With this form of the potential, σ is the equilibrium inter-atomic distance between two particles and −ε is the corresponding potential value. Temperatures and energies are measured in units of ε.   We will not describe the details of

the work, though their Fig.1 is presented as Fig.7.5.4. It shows how each
layer is formed as the number of introduced atoms increases. (Note that one
layer can accommodate 224 atoms.) The upper figure corresponds to a case of
relatively high substrate temperature $T_s=0.4$. (The melting temperature of LJ
crystal is about 0.7.) In this case, the mobility of particles is large and the
growth is into fully completed layers. On the other hand, at $T_s=0$ (the lower
figure) the number of particles in each layer is always less than 224 and it
decreases with increasing height. Therefore an appreciable amount of vacancies,
voids, and grain boundaries exist. Nevertheless, it was shown that the

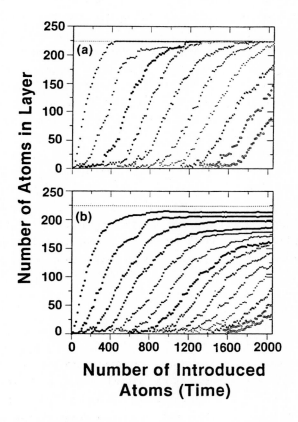

*Figure 7.5.4 Molecular-dynamics simulation of epitaxial
growth (from Ref.95). The number of particles in successive
layers (from left to right) is shown as a function of the
number of introduced atoms. Values on the abscissa are
proportional to the simulation time. The dashed horizontal
line indicates the number of particles in a fully completed
close-packed layer. (a) is at $T_s=0.4$ and (b) at $T_s=0$, with
T_s being the substrate temperature.*

adsorbate consists of well-defined layers and the in-plane atomic arrangement has a good hexagonal pattern even in the fifth plane from the substrate.

An extension of this sort of simulation to the superlattice growth seems to be rather straightforward and extensive studies will be done in the near future. An analysis should be done of the supermodulus effect based on the structure obtained by such a simulation. It is also interesting to study how the asymmetry of the atomic configuration at the two interfaces, A on B and B on A, comes about. Of course, such an asymmetry may be partly explained by the fact that at the initial stage of deposition of A on B, the system may be regarded as an alloy of dilute A atoms in bulk B and vice versa, and the solubility of A into B may generally be different from that of B into A in the thermal equilibrium. However, the asymmetry in the superlattices is sometimes much more drastic than what we expect from the equilibrium phase diagrams. One of the most drastic asymmetries observed so far is the case of Fe/Mn(96). When Mn is deposited on Fe, the concentration modulation is fairly sharp, though it is very broad in the opposite case. Actually in the latter broad transient region, formation of a new alloy phase was concluded experimentally(96). Theoretical studies on this subject are highly desirable.

A fundamental difficulty in the molecular-dynamics study is the determination of interatomic potentials. Simple LJ-type potentials will certainly have limitations to their applicability to real systems and extensive studies on the interatomic potentials are now in progress(97). Furthermore, a new technique for applying the density functional method to molecular dynamics was proposed recently(98), which may enable us to deal with a variety of systems on a sound basis.

7.6 SUMMARY

We have surveyed some basic aspects of magnetic, elastic, and structural properties of metallic superlattices. As was stated in Chapter 1, the number of publications in this field is still growing rapidly and a lot of systems have been studied experimentally. On the other hand, so long as the theoretical works are concerned, only a few unified-theory-type works are available. Theoretical works have been done mostly to analyze specific problems. If we try to analyze the observed properties on the atomistic level, we will have difficulty obtaining detailed information about the atomic configuration. This is an inevitable aspect of the work in this field. Nevertheless, there have been many examples of fruitful collaboration between theory and experiment.

Considering the rapid progress of computers, we believe that computer simulations such as the molecular dynamics and the Monte Carlo method will play even more significant roles in the future. The most difficult point at present with these simulations is that we do not have a good prescription for determining the interatomic potentials to describe specific systems. A new technique(98), which combines the density functional electronic structure calculation with the molecular dynamics, seems to be promising in this context.

There are some important topics that we have not covered in this chapter. Theoretical(99, 100) and experimental(101,102) works on the magneto-static modes in superlattices composed of magnetic and non-magnetic layers are such examples. Very beautiful experiments have recently been done for Fe/Pd and Fe/W(102), which are the first systematic studies on the collective spin waves in the superlattices. Another example is the superlattice containing rare-earth elements such as Gd/Y(103,104) and Dy/Y(105), where the RKKY interaction produces a long range magnetic interaction between magnetic layers mediated by the nonmagnetic Y layers. A fairly long range magnetic interlayer interaction was also observed in Fe/Cr(106). The possibility of having a variety of magnetic orderings was discussed by Hinchey and Mills(107).

Acknowledgement

I am indebted to Dr. N. Hamada for collaboration on the subject presented in Sec. 7.2 (1) and also for valuable comments on the local-environment effects. Discussions with Dr. T. Oguchi on the subject of Sec. 7.3 were also very useful. I would like to express my sincere thanks to Miss Oto for her help in preparing this manuscript.

244

REFERENCES

(1) A review of surface magnetism : M. Weinert, A. J. Freeman, S. Ohnishi and J. W. Davenport, J. Appl. Phys. 57 (1985) 3641.
(2) Arguments by V. Heine on the magnetic moments of two inequivalent iron sites in Fe_3Al: V. Heine, in "Solid State Physics", vol.35, ed. by H. Ehrenreich, F. Seitz and D. Turnbull (Academic Press, 1980) p.1.
(3) N. Hamada, K. Terakura and A. Yanase, J. Phys. F 14 (1984) 2371.
(4) N. Hosoito, K. Kawaguchi, T. Shinjo, T. Takada and Y. Endoh, J. Phys. Soc. Jpn. 53 (1984) 2659.
(5) Y. Endoh, K. Kawaguchi, N. Hosoito, T. Shinjo, T. Takada, Y. Fujii and T. Ohnishi, J. Phys. Soc. Jpn. 53 (1984) 3481.
(6) K. Takanashi, H. Yasuoka, K. Kawaguchi, N. Hosoito and T. Shinjo, J. Phys. Soc. Jpn. 53 (1984) 4315.
(7) H. K. Wong, H. Q. Yang, B. Y. Jin, Y. H. Shen, W. Z. Cao and J. B. Ketterson, J. Appl. Phys. 55 (1984) 2494.
(8) N. K. Jaggi, L. H. Schwartz, H. K. Wong and J. B. Ketterson, J. Magn. Magn. Mater. 49 (1985) 1.
(9) H. K. Wong, H. Q. Yang, J. E. Hilliard and J. B. Ketterson, J. Appl. Phys. 57 (1985) 3660.
(10) N. Hamada, K. Terakura, K. Takanashi and H. Yasuoka, J. Phys. F 15 (1985) 835.
(11) M. E. Elzain, D. E. Ellis and D. Guenzburger, Phys. Rev. B 34 (1986) 1430.
(12) N. Hamada and H. Miwa, Prog. Theor. Phys. 59 (1978) 1045.
(13) N. Hamada, J. Phys. Soc. Jpn. 50 (1981) 77.
(14) N. Hamada, j. Phys. Soc. Jpn. 53 (1984) 3172.
(15) I. Mirebeau, M. C. Cadeville, G. Parette and I. A. Campbell, J. Phys. 12 (1982) 25.
(16) J. F. Janak, Phys. Rev. B 20 (1979) 2206.
(17) B. J. Thaler, J. B. Ketterson and J. E. Hilliard, Phys. Rev. Lett. 41 (1978) 336.
(18) T. Jarlborg and A. J. Freeman, Phys. Rev. Lett. 45 (1980) 653.
(19) T. Jarlborg and A. J. Freeman, J. Appl. Phys. 53 (1982) 8041.
(20) A. J. Freeman, J. Xu and T. Jarlborg, J. Magn. Magn. Mater. 31-34 (1983) 909.
(21) J. Tersoff and L. M. Falicov, Phys. Rev. B 26 (1982) 6186.
(22) L. R. Sill, M. M. Brodsky, S. Bowen and H. C. Hamaker, J. Appl. Phys. 57 (1985) 3663.
(23) N. K. Flevaris, J. B. Ketterson and J. E. Hilliard, J. Appl. Phys. 53 (1982) 8046.
(24) E. M. Gyorgy, D. B. McWhan, J. F. Dillon, Jr., L. R. Walker and J. V. Waszczak, Phys. Rev. B 25 (1982) 6739.
(25) E. M. Gyorgy, D. B. McWhan, J. F. Dillon, Jr., L. R. Walker, J. V. Waszczak, D. P. Musser and R. H. Willens, J. Magn. Magn. Mater. 31-34 (1983) 915.
(26) G. Shirane and W. J. Takei, J. Phys. Soc. Jpn. 17, Suppl. B-III (1962) 35.
(27) Y. Teraoka and J. Kanamori, Transition Metal 1977 (Inst. Phys. Conf. Ser. 39) p.588.
(28) G. Allan, Surf. Sci. 74 (1978) 79.
(29) G. Allan, Phys. Rev. B 19 (1979) 4774.
(30) G. Allan, Surf. Sci. Rep. 1 (1981) 121.
(31) D. R. Grempel, Phys. Rev. B 24 (1981) 3928.
(32) R. L. Stratonovich, Sov. Phys. Dokl. 2 (1957) 416.
(33) J. Hubbard, Phys. Rev. Lett. 3 (1959) 77.
(34) J. Hubbard, Phys. Rev. B 19 (1979) 2626.
(35) H. Hasegawa, J. Phys. Soc. Jpn. 46 (1979) 1504.
(36) H. Hasegawa, J. Phys. Soc. Jpn. 49 (1981) 178.
(37) L. E. Klebanoff, S. W. Robey, G. Liu and D. A. Shirley, Phys. Rev. B 30 (1984) 1048.
(38) L. E. Klebanoff, R. H. Victora, L. M. Falicov and D. A. Shirley, Phys. Rev.

B 32 (1985) 1997.
(39) L. E. Klebanoff, S. W. Robey, G. Liu and D. A. Shirley, Phys. Rev. B 31 (1985) 6379.
(40) L. E. Klebanoff, S. W. Robey, G. Liu and D. A. Shirley, J. Magn. Magn. Mater. 728 (1986) 728.
(41) R. H. Victora and L. M. Falicov, Phys. Rev. B 31 (1985) 7335.
(42) H. Hasegawa, J. Phys. F 16 (1986) 1555.
(43) As a review article, J. E. Inglesfield, Prog. Surf. Sci. 20 (1985) 105.
(44) A. Kallel and F. de Bergevin, Solid St. Comm. 5 (1967) 955.
(45) W. Koster, F. Wechtel and K. Grube, Z. Metalk. 54 (1963) 393.
(46) S. Araj, N. L. Reeves and E. E. Anderson, J. Appl. Phys. 42 (1971) 1691.
(47) S. Matsuo and I. Nishida, J. Phys. Soc. Jpn. 49 (1980) 1005.
(48) S. Ohnishi, A. J. Freeman and M. Weinert, Phys. Rev. B 28 (1983) 6741.
(49) E. Wimmer, A. J. Freeman and H. Krakauer, Phys. Rev. B 30 (1984) 3113.
(50) U. Gradmann, G. Waller, R. Fedder and E. Tamura, J. Magn. Magn. Mater. 31-34 (1983) 883.
(51) A. M. Turner and J. L. Erskine, Phys. Rev. B 28 (1983) 5628.
(52) J. L. Erskine, Phys. Rev. Lett. 45 (1980) 1446.
(53) E. Kisker, K. Schröder, W. Gudat and M. Campagna, Phys. Rev. B 31 (1985) 329.
(54) T. Aldred, J. Phys. C 1 (1968) 1103.
(55) J. Crangle and M. J. C. Martin, Phil. Mag. 4 (1955) 1006.
(56) K. Terakura, Physica 91 B+C (1977) 162.
(57) C. L. Fu, A. J. Freeman and T. Oguchi, Phys. Rev. Lett. 54 (1985) 2700.
(58) T. Oguchi and A. J. Freeman, J. Magn. Magn. Mater. 54-57 (1986) 797.
(59) T. Oguchi and A. J. Freeman, in preparation.
(60) S. Ohnishi, M. Weinert and A. J. Freeman, Phys. Rev. B 30 (1984) 36.
(61) R. Richter, J. G. Gay and J. R. Smith, Phys. Rev. Lett. 54 (1985) 2704.
(62) T. Shinjo, K. Kawaguchi, T. Yamamoto, N. Hosoito and T. Takada, Solid State Commun. 52 (1984) 257 and K. Kawaguchi, R. Yamamoto, N. Hosoito, T. Shinjo and T. Takada, J. Phys. Soc. Jpn. 55 (1986) 2375.
(63) T. Shinjo, N. Nakayama, I. Moritani and Y. Endoh, J. Phys. Soc. Jpn. 55 (1986) 2512.
(64) P. F. Carcia, A. D. Meinhaldt and A. Suna, Appl. Phys. Lett. 47 (1985) 178.
(65) B. T. Jonker, K. H. Walker, E. Kisker, G. A. Prinz and C. Carbone, Phys. Rev. Lett. 57 (1986) 142.
(66) J. G. Gay and R. Richter, Phys. Rev. Lett. 56 (1986) 2728.
(67) G. Dublon, M. P. Rosenblum and W. T. Vetterling, IEEE Trans. Magn. Mag-16 (1980) 1126.
(68) U. Gradmann, J. Magn. Magn. Mater. 54-57 (1986) 733.
(69) S. D. Bader, E. R. Moog and P. Grünberg, J. Magn. Magn. Mater. 53 (1986) L295.
(70) L. Néel, J. Phys. Radium 15 (1954) 225.
(71) A. J. Bennett and B. R. Cooper, Phys. Rev. B 3 (1971) 1642.
(72) H. Takayama, K. P. Bohnen and P. Fulde, Phys. Rev. B 14 (1976) 2287.
(73) K. Terakura, J. Phys. C 11 (1978) 469.
(74) K. Terakura, N. Hamada, T. Oguchi and T. Asada, J. Phys. F 12 (1982) 1661.
(75) W. M. C. Yang, T. Tsakalakos and J. E. Hilliard, J. Appl. Phys. 48 (1977) 876.
(76) T. Tsakalakos and J. E. Hilliard, J. Appl. Phys. 54 (1983) 734.
(77) G. E. Henein and J. E. Hilliard, J. Appl. Phys. 54 (1983) 728.
(78) A. Kueny, M. Grimsditch, K. Miyano, I. Banerjee, C. M. Falco and I. K. Schuller, Phys. Rev. Lett. 48 (1982) 166.
(79) M. R. Khan, C. S. L. Chun, G. P. Felcher, M. Grimsditch, A. Kueny, C. M. Falco and I. K. Schuller, Phys. Rev. B 27 (1983) 7186.
(80) R. Danner, R. P. Huebener, C. S. L. Chun, M. Grimsditch and I. K. Schuller, Phys. Rev. B 33 (1986) 3696.
(81) I. K. Schuller, Proc. of the IEEE Ultrasonic Symposium (San Francisco, 1985), p.1093.
(82) W. E. Pickett, J. Phys. F 12 (1982) 2195.
(83) T. B. Wu, J. Appl. Phys. 53 (1982) 5265.

(84) D. Baral and J. E. Hilliard, Appl. Phys. Lett. 41 (1982) 156.
(85) D. Baral, J. B. Ketterson and J. E. Hilliard, in "Modulated Structure Materials" ed. by T. Tsakalakos (NATO ASI Series, Series E: Applied Sciences, No.83, 1984) p.465.
(86) P. C. Clapp, p.455 in the book of ref.85.
(87) T. Tsakalakos and A. F. Jankowski, p.387 in the book of ref.85.
(88) M. Imafuku, Y. Sasajima, R. Yamamoto and M. Doyama, J. Phys. F 16 (1986) 823.
(89) I. K. Schuller and A. Rahman, Phys. Rev. Lett. 50 (1983) 1377.
(90) B. W. Dodson, Phys. Rev. B30 (1984) 3545.
(91) R. Ramirez, A. Rahman and I. K. Schuller, Phys. Rev. B30 (1984) 6208.
(92) A. R. Verma and P. Krishna, "Polymorphism and Polytypism in Crystals" (Wiley, New York, 1966).
(93) J. E. Cunningham and C. P. Flynn, J. Phys. F15 (1985) L221.
(94) A. C. Redfield and A. M. Zangwill, Phys. Rev. B34 (1986) 1378.
(95) M. Schneider, A. Rahman and I. K. Schuller, Phys. Rev. Lett. 55 (1985) 604.
(96) K. Takanashi, H. Yasuoka, N. Nakayama, T. Katamoto and T. Shinjo, J. Phys. Soc. Jpn. 55 (1986) 2357 T. Shinjo, Hyperfine Interactions 27 (1986) 193. N. Nakayama, T. Katamoto and T. Shinjo, in preparation.
(97) R. Biswas and D. R. Hamann, Phys. Rev. Lett. 55 (1985) 2001.
(98) R. Car and M. Parrinello, Phys. Rev. Lett. 55 (1985) 2471.
(99) R. E. Camley, T. S. Rahman and D. L. Mills, Phys. Rev. B 27 (1983) 261.
(100) P. Grünberg and K. Mika, Phys. Rev. B 27 (1983) 2955.
(101) M. Grimsditch, M. R. Khan, A. Kueny and I. K. Schuller, Phys. Rev. Lett. 51 (1984) 498.
(102) B. Hillebrands, P. Baumgart, R. Mock, G. Güntherodt, A. Boufelfel and C. M. Falco, to be published in Phys. Rev. B.
(103) J. Kwo, E. M. Gyorgy, D. B. McWhan, M. Hong, F. J. DiSalvo, C. Vettier and J. E. Bower, Phys. Rev. Lett. 55 (1985) 1402.
(104) C. F. Majkrzak, J. W. Cable, J. Kwo, M. Hong, D. B. McWhan, Y. Yafet, J. V. Waszczak and C. Vettier, Phys. Rev. Lett. 56 (1986) 2700.
(105) M. B. Salamon, S. Sinha, J. J. Rhyne, J. E. Cunningham, R. W. Erwin, J. Borchers and C. P. Flynn, Phys. Rev. Lett. 56 (1986) 259.
(106) P. Grünberg, R. Schreiber, Y. Pang, M. B. Brodsky and H. Sowers, Phys. Rev. Lett. 57 (1986) 2442.
(107) L. L. Hinchey and D. L. Mills, Phys. Rev. B 33 (1986) 3329.

Appendix

I. Alphabetical Classification of Constituents
[Combination of elements, Method of preparation(VD and SP mean vacuum
deposition and sputtering, respectively), Main subject or measurement, and
Code number of the reference]

Ag / Au	VD	interdiffusion	69C1
Ag / Co	VD	magnetization, ferromagnetic resonance	85T2
Ag / In	VD	superconductivity, Hc	79G1
Ag / Ni	VD	magnetization, ferromagnetic resonance	86K9
Ag / Pb	VD	X-ray diffraction	83J1
	VD	annealing effect	83J2
	VD	tunneling	84J1
Ag / Pd	VD	elastic modulus	83H1
	VD	interdiffusion	84H1
Ag / Sn	VD	superconductivity, Tc	79G1
Ag / V	VD	superconductivity, Hc	86K1
Al / Ge	VD	superconductivity, Tc, Hc	78H1
Al / Nb	SP	X-ray diffraction	83M1
Au / Co	SP	Kerr rotation	86K7
Au / Cu	VD	interdiffusion	77P1
	VD	elastic modulus	83H1
Au / Fe	SP	Kerr rotation	86K7
Au / Ge	VD	superconductivity, Tc, Hc	85A1
Au / Ni	VD	elasticity	77Y1
Au / Pd	SP	X-ray diffraction, electric resistivity	83C2
C / Re	VD	X-ray mirror	81G1
C / Ta	VD	X-ray mirror	81G1
C / W	VD	X-ray mirror	81G1
	SP	X-ray diffraction, electron microscopy	85T1
Co / Cu	SP	Kerr rotation	86K7
Co / Cr	SP	magnetization bcc-Co	83W1
	VD	magnetization, ferromagnetic resonance	85T2
	VD	magnetization	86S9
Co / Gd	SP	magnetization	85W2
Co / Mn	VD	magnetization	82S1

	VD	spin glass, magnetization	84S6
	VD	ferromagnetic resonance, magnetization	85S1
	VD	magnetization, spin glass	85Z1
	VD	nuclear magnetic resonance	86D1
	VD	magnetization	86S1
Co / Nb	VD	ferromagnetic resonance	84K3
	VD	ferromagnetic resonance, magnetization	85K2
Co / Pd	SP	magnetization	85C2
	SP	magnetization, electric resistivity	85C3
Co / Sb	VD	nuclear magnetic resonance	84T2
	VD	X-ray diffraction	86N1
Co(Nb)/Co(Ti)	SP	magnetization	86K2
Cr / Fe	VD	magnetization	86S4
Cu / Fe	VD	hardness	80B1
	VD	high temperature creep	81M2
	SP	Mössbauer spectroscopy	85D2
	SP	Mössbauer spectroscopy	85N1
	SP	Kerr rotation	86K7
Cu / Nb	SP	X-ray diffraction	80S1
	SP	X-ray diffraction, electric resistivity	81L1
	SP	superconductivity, Tc	82B1
	SP	Brillouin scattering	82K1
	SP	structure superconductivity, Tc, Hc	82S2
	SP	superconductivity, Hc	83B1
	SP	tunneling	83Y1
	SP	superconductivity, Tc, Hc	84B2
	SP	superconductivity, Hc	84C1
	SP	superconductivity, Tc, Hc	85H3
	SP	nuclear magnetic resonance(Cu)	85Y2
Cu / Ni	VD	ferromagnetic resonance	78T1
	VD	hardness	80B1
	VD	neutron diffraction	80F1
	VD	magnetization	80G1
	VD	electron microscopy	80N1
	VD	ferromagnetic resonance	81D1
	VD	X-ray diffraction	81F1
	VD	high temperature creep	81M2
	VD	magnetization	81Z1
	VD	magnetization	82F1
	VD	magnetization	82F2

	VD	magnetization	82F3
	VD	magnetization	82G1
	VD	Auger electron spectroscopy, interdiffusion	82R2
	VD	magnetization, Tc	82Z1
	VD	magnetization and structure	83G2
	VD	elastic modulus	83T1
	VD	interdiffusion	84T3
	VD	Young's & torsion moduli	85B1
Cu/Ni(Fe)	SP	ferromagnetic resonance	83S6
Cu / Pd	VD	interdiffusion	69P1
	VD	elasticity	77Y1
Cu / Ti	SP	Auger electron spectroscopy	86H2
Dy / Y	MBE	magnetization, neutron diffraction	86S2
	MBE	magnetization	86S3
Er / Nb	SP	magnetism & superconductivity	85G1
Fe / Gd	VD	magnetization	85M3
	VD	magnetization, Mössbauer spectroscopy	85U1
	VD	magnetization	86M2
Fe / Ge	SP	neutron diffraction	85M1
	SP	neutron diffraction	86M4
Fe / Mg	VD	interdiffusion	67D1
	VD	X-ray diffraction	84S4
	VD	Mössbauer spectroscopy	84S5
	VD	Mössbauer spectroscopy	85S4
	VD	X-ray diffraction	86F1
	VD	magnetization, Mössbauer spectroscopy	86S8
	VD	magnetization, Mössbauer spectroscopy	86K5
Fe / Mn	VD	nuclear magnetic resonance	86T3
Fe / Pd	VD	magnetization, Mössbauer spectroscopy	80D1
	SP	Brillouin scattering	86B1
	SP	Brillouin scattering	86B2
	SP	Brillouin scattering	86H1
	SP	Brillouin scattering	86H4
Fe / Sb	VD	neutron diffraction	83E1
	VD	Mössbauer spectroscopy(Sb-121 & Fe-57)	83F1
	VD	Mössbauer spectroscopy, neutron diffraction, FMR	83S4
	VD	Mössbauer spectroscopy	83S5
Fe(C)/Si	SP	diffusion & magnetization	83K1
	SP	magnetization	84K1
Fe / Sm	VD	magnetization, Mössbauer spectroscopy	86U2

Fe / Tb	SP	magnetization	86S3
Fe / V	VD	X-ray diffraction	84E1
	VD	magnetization, Mössbauer spectr., neutron diffraction	84H2
	VD	Mössbauer spectroscopy	84S3
	VD	nuclear magnetic resonance	84T4
	VD	X-ray diffraction, magnetization, superconductivity	84W1
	VD	Mössbauer spectroscopy	85J1
	VD	Mössbauer spectroscopy	85J2
	VD	Mössbauer spectroscopy	85S2
	VD	magnetism & superconductivity	85W1
	VD	nuclear magnetic resonance	86T2
Fe / Y	VD	magnetization	86M2
Fe / W	SP	Brillouin scattering	86H4
Gd / Y	MBE	X-ray diffraction, magnetization	85K3
	MBE	magnetization	86K3
	MBE	magnetization	86K4
	MBE	magnetic X-ray scattering	86M1
	MBE	neutron diffraction	86M3
	MBE	magnetic X-ray scattering	86V1
	MBE	electron diffraction	86K6
Ge / Nb	SP	superconductivity, Hc	80R1
	SP	superconductivity, Hc	82R1
Ge/Nb(Ti)	SP	superconductivity, Hc	85J3
Ir / Ru	VD	superconductivity	86C2
Lu / Nb	SP	superconductivity	85G1
Lu / Tm	SP	magnetization	85L1
Mn / Ni	VD	magnetization	82S1
	VD	magnetization, spin glass	84S6
	VD	magnetization, spin glass	85V3
	VD	magnetization, spin glass	85Z1
Mo / Ni	SP	Brillouin scattering	83G1
	SP	structure, elasticity, electric resistivity	83K2
	SP	elasticity	83S2
	SP	Brillouin scattering	84K2
	SP	magnetization, Brillouin scattering	84S2
	SP	electric resistivity	84U1
	SP	electric resistivity	85C1
	SP	ferromagnetic resonance	85P1
	SP	superconductivity, Hc	86U1
	SP	superconductivity, Hc	86U3

	SP	X-ray diffraction, electron diffraction	86Z1
Mn / Sb	VD	X-ray diffraction, magnetization	86N2
Mo/Sb$_3$N$_4$	SP	electrical resistivity, magnetoresistance	86N3
Mo / Sb	VD	superconductivity, Tc	86A1
Mo / Si	SP	superconductivity, Tc, Hc. Jc	83I1
Mo / V	SP	superconductivity, Tc	85K4
	SP	superconductivity, Tc	85K5
	SP	quasiperiodicity, superconductivity, Hc	86K8
Nb(Ge)/Nb(Si)	SP	electric resistivity, superconductivity	85Y1
Nb / Pd	SP	hydrogen solubility	85M6
Nb / Ta	MBE	X-ray diffraction, electron microscopy	81D2
	MBE	hydrogen solubility, induced strain	85M2
Nb / Ti	SP	superconductivity, Tc	81Z2
Nb / Tm	SP	magnetism & superconductivity	85G1
Nb / V	VD	X-ray diffraction	86L1
Nb / Zr	SP	specific heat	84B1
	SP	EXAFS	84C2
	SP	structure & superconductivity, Tc	84L1
Ni(Fe)/Ni	SP	ferromagnetic resonance	85S3
Ni / Mo	SP	magnetization	86C1
Ni / Pd	VD	magnetization	82F2
Ni / V	SP	superconductivity, Hc, & magnetism	85H3
	SP	Brillouin scattering	86D2
Re(W)/C	VD	X-ray mirror	79H2
Ti/Si$_3$N$_4$	SP	electrical resistivity, magnetoresistance	86N3

II. List of Publications

[code number, author(s), journal, volume (year) pages, and title.
R and T attached to the code numbers mean review and theoretical articles,
respectively.]

67D1 J.B. Dinklage, J. Appl. Phys., 38(1967)3781-5, X-ray Diffraction by
      Multilayered Thin-Film Structures and Their Diffusion

69C1 H.E. Cook and J.E. Hilliard, J. Appl. Phys., 40(1969)2191-8, Effect of
      Gradient Energy on Diffusion in Gold-Silver Alloys

69P1 E.M. Philofsky and J.E. Hilliard, J. Appl. Phys., 40(1969)2198-205,
      Effect of Coherency Strains on Diffusion in Copper-Palladium Alloys

77P1 W.M. Paulson and J.E. Hilliard, J. Appl. Phys., 48(1977)2117-23,
      Interdiffusion in Composition-Modulated Copper-Gold Thin Films

77Y1 W.M.C. Yang, T. Tsakalakos, and J.E. Hilliard, J. Appl. Phys., 48(1977)
      876-9, Enhanced Elastic Modulus in Composition-Modulated Gold-Nickel and
      Copper-Palladium Foil

78H1 T.W. Haywood and D.G. Ast, Phys. Rev. B18(1978)2225-36, Critical Fields
      of Multilayerd Films of Al and Ge

78T1 B.J. Thaler, J.B. Ketterson and J.E. Hilliard, Phys. Rev. Letters,
      41(1978)336-9, Enhanced Magnetization Density of a Compositionally
      Modulated CuNi Thin Film

79G1 C.G. Granqvist and T. Claeson, Solid State Commun., 32(1979)531-5,
      Superconducting Transition Temperatures of Vapour Quenched Ag-In and
      Ag-Sn Multilayers

79H1R J.E. Hilliard, AIP Conf. Proc. No.53(1979)407-16, Artificial Layer
      structures and their Properties

79H2 R.P. Haelbich, A. Segmüller and E. Spiller, Appl. Phys. Lett.,
      34(1979)184-6, Smooth Multilayer Films Suitable for X-Ray Mirrors

80B1 R.F. Bunshah, R. Nimmagadda, H.J. Doerr, B.A. Movchan, N.I. Grechanuk
      and E.V. Dabizha, Thin Solid Films, 72(1980)261-75, Structure and
      Property Relationships in Microlaminate Ni-Cu and Fe-Cu Condensates

80D1 G. Dublon, M.P. Rosenblum and W.T. Vetterling, IEEE Trans., Mag-16
      (1980)1126-8, Magnetic Properties of Compositionally Modulated
      Pd/Fe Films

80F1 G.P. Felcher, J.W. Cable, J.Q. Zheng, J.B. Ketterson and J.E. Hilliard:
      J. Magn. & Magn. Mater., 21(1980)L198-202, Neutron Diffraction Analysis
      of a Compositionally Modulated Alloy of Nickel-Copper

80G1 E.M. Gyorgy, J.F. Dillon Jr., D.B. McWhan, L.W. Rupp Jr., L.R. Testardi
      and P.J. Flanders, Phys. Rev. Letters, 45(1980)57-60, Magnetic
      Properties of Compositionally Modulated Cu-Ni Thin Films

80J1T T. Jarlborg and A.J. Freeman, Phys. Rev. Letters, 45(1980)653-6,
      Electronic Structure and Magnetism of CuNi Coherent Modulated Structures

80N1 S. Nakahara, R.J. Schutz and L.R. Testardi, Thin Solid Films, 72(1980)
      277-84, Transmission Electron Microscope Study of Thin Cu-Ni Laminates

Obtained by Vapor Deposition

80R1 S.T. Ruggiero, T.W. Barbee, Jr. and M.R. Beasley, Phys. Rev. Lett., 45(1980)1299-302, Superconductivity in Quasi-Two-dimensional Layered Composites

80S1 I.K. Schuller, Phys. Rev. Lett., 44(1980)1597-1600, New Class of Layered Materials

80S2 E. Spiller, A. Segmüller, J. Rife and R.P. Haelbich, Appl. Phys. Lett., 37(1980)1048-50, Controlled Fabrication of Multilayer Soft-X-Ray Mirrors

80S3 M. Sato, K. Abe, Y. Endoh and J. Hayter, J. Phys., C3(1980)3563-76, Magnetization of Ferromagnetic Metals at the Interface of Other Materials

80W1T R.M. White and C. Herring, Phys. Rev. B22(1980)1465-6, Magnetic Resonance in Multilayer Films

81D1 J.F. Dillon Jr., E.M. Gyorgy, L.W. Rupp Jr., Y. Yafet and L.R. Testardi, J. Appl. Phys., 52(1981)2256-8, Ferromagnetic Resonance in Compositionally Modulated Cu-Ni Thin Films

81D2 S.M. Durbin, J.E. Cunningham, M.E. Mochel and C.P. Flynn, J. Phys. F; 11(1981)L223-6, Nb-Ta Metal Superlattices

81F1 N.K. Flevaris, D. Baral, J.E. Hilliard and J.B. Ketterson, Appl. Phys. Lett., 38(1981)992-4, A Note on Compositionally Modulated Cu-Ni Films with Lattice-Commensurate Wavelengths

81G1 S.V. Gaponov, S.A. Gusev, B.M. Luskin, N.N. Salashchenko and E.S. Gluskin, Optics Commun. 38(1981)7-9, Long-WaveX-Ray Radiation Mirrors

81J1T T. Jarlborg and A.J. Freeman, J. Appl. Phys., 52(1981)1622-3, Electronic Structure and Magnetism of CuNi Coherent Modulated Structures

81K1 J. Koehler, Phys. Rev. B23(1981)1753-60, Layered Solids. New Crystalline Materials

81L1 W.P. Lowe, T.W. Barbee Jr., T.H. Geballe and D.B. McWhan, Phys. Rev. B24 (1981)6193-6, X-ray Scattering from Multilayers of NbCu

81L2T P. LEE, Optics Commun., 37(1981)159-64, X-Ray Diffraction in Multilayers

81M1 K.E. Meyer, G.P. Felcher, S.K. Sinha and I.K. Schuller, J. Appl. Phys., 52(1981)6608-10, Models of Diffraction from Layered Ultrathin Coherent Structures

81M2 B.A. Movchan, E.V. Dabizha, R.F. Bunshah and R.R. Nimmagadda, Thin Solid Films, 83(1981)21-6, High Temperature Creep Study of Fe/Cu and Ni/Cu Multilayer Condensates

81R1T H. Raffy and E. Guyon, Physica, 108B(1981)947-8, Dependence of Critical Current and Fields of Periodically Modulated Superconducting Alloys on Mudulation Amplitude

81T1 L.R. Testardi, R.H. Willens, J.T. Krause, D.B. McWhan and S. Nakahara, J. Appl. Phys., 52(1981)510-1, Enhanced Elastic Moduli in Cu-Ni Films with Compositional Modulation

81U1R   J.H. Underwood and T.W. Barbee, Jr., Appl. Opt., 20(1981)3027-34,
        Layered Synthetic Microstuctures as Bragg Diffractors for X Rays and
        Extreme Ultraviolet: Theory and Predicted Performance

81Z1   J.Q. Zheng, C.M. Falco, J.B. Ketterson and I.K. Schuller: Appl. Phys.
        Letters, 38(1981)424-6, Magnetization of Compositionally Modulated CuNi
        Films

81Z2   J.Q. Zeng, J.B. Ketterson, C.M. Falco and I.K. Schuller, Physica,
        108B(1981)945-6, Superconducting and Transport Properties of NbTi
        Layered Metals

82B1   I. Banerjee, Q.S. Yang, C.M. Falco and I.K. Schuller, Solid State
        Commun., 41(1982)805-8, Superconductivity of Nb/Cu Superlattices

82C1T   A. Caillé, M. Banville, P.D. Loly and M.J. Zuckermann, Solid State
        Commun., 41(1982)119-22, The Crossover from Two-Dimensional to Three-
        Dimentional Plasmon Behavior in Layered Systems

82F1   N.K. Flevaris, J.B. Ketterson and J.E. Hilliard, J. Appl. Phys.,
        53(1982)1997-2001, Magnetic Properties of Compositionally Modulated
        Thin Films

82F2   N.K. Flevaris, J.B. Ketterson and J.E. Hilliard, J. Appl. Phys.,
        53(1982)2439-44, Static and Dynamic Magnetic Studies of Compositionally
        Modulated Thin Films

82F3   N.K. Flevaris, J.B. Ketterson and J.E. Hilliard, J. Appl. Phys.,
        53(1982)8046-51, Magnetic Properties of Compositionally Modulated Thin
        Films

82F4T   A.J. Freeman, D.S. Wang and H. Krakauer, J. Appl. Phys., 53(1982)
        1997-2001, Magnetism of Surfaces and Interfaces

82G1   E.M. Gyorgy, D.B. McWhan, J.F. Dillon Jr., L.R. Walker and
        J.V. Waszczak, Phys. Rev. B25(1982)6739-47, Magnetic Behavior and
        Structure of Compositionally Modulated Cu-Ni Thin Films

82H1   N. Hosoito, K. Kawaguchi, T. Shinjo and T. Takada, J. Phys. Soc. Jpn.,
        51(1982)2701-2, Monatomic Fe Layer Sandwiched in Sb Layers

82J1T   T. Jarlborg and A.J. Freeman, J. Appl. Phys., 53(1982)8041-5, Magnetism
        of Metallic Superlattices

82J2   W.R. Jones, J. Appl. Phys., 53(1982)2442-4, Computer Analysis of Arrott
        Plots of Compositionally Modulated Materials

82K1   A. Kueny, M. Grimsditch, K. Miyano, I. Banerjee, C.M. Falco and
        I.K. Schuller, Phys. Rev. Lett., 48(1982)166-70, Anomalous Behavior of
        Surface Acoustic Waves in Cu/Nb Superlattices

82L1   P. LEE, Opt. Commun., 42(1982)199, Application of the WKB Method to
        X-Ray Diffraction by One Dimensional Periodic Structures

82R1   S.T. Ruggiero, T.W. Barbee Jr., and M.R. Beasley, Phys. Rev. B26(1982)
        4894-908, Superconducting Properties of Nb/Ge Metal Semiconductor
        Multilayers

82R2   K. Röll and W. Reill, Thin Solid Films, 89(1982)221-4, A Study of
        Interdiffusion in Multilayer Cu/Ni Films by Auger Electron Depth
        Profiling

82S1   M.B. Stearns, J. Appl. Phys., 53(1982)2436-8, Magnetization and
        Structure of Mn-Ni and Mn-Co Layered Magnetic Thin Films

82S2   I.K Schuller and C.M. Falco, Thin Solid Films, 90(1982)221,
        Superconductivity and Magnetism in Metallic Superlattices

82Z1   J.Q. Zheng, J.B. Ketterson, C.M. Falco, I.K. Schuller, J. Appl. Phys.,
        53(1982)3150, The Magnetization and Temperature of Compositionally
        Modulated Cu/Ni Films

83B1   I. Banerjee, Q.S. Yang, C.M. Falco and I.K. Schuller, Phys. Rev. B28
        (1983)5037-40, Anisotropic Critical Fields in Superconducting
        Superlattices

83C1T   R.E. Camley, T.S. Rahman and D.L. Mills, Phys. Rev. B27(1983)261-77,
        Magnetic Excitations in Layered Media: Spin Waves and the
        Light-Scattering Spectrum

83C2   P.F. Carcia and A. Suna, J. Appl. Phys., 54(1983)2000-5, Properties of
        Pd/Au Thin Film Layered Structures

83C3   W.K. Chu, C.K. Pan and C.A. Chang, Phys. Rev. B28(1983)4033-6,
        Superlattice Interface and Lattice Strain Measurement by Ion Channeling

83E1   Y. Endoh, H. Ono, N. Hosoito and T. Shinjo, J. Magn. & Magn. Mater.,
        31-34(1983)881-2, Magnetism of Fe Interface Studied by Neutron
        Diffraction

83E2   Y. Endoh, N. Hosoito and T. Shinjo, J. Magn. & Magn. Mater., 35(1983)
        93-8, Application of Neutron Diffraction to the Study of Interface
        Magnetization on Thin Films with Artificial Superlattices

83F1   J.M. Friedt, N. Hosoito, K. Kawaguchi and T. Shinjo, J. Magn. & Magn.
        Mater., 35(1983)136-8, Interface Magnetism in Fe-Sb Multilayer Films
        from ^{121}Sb and ^{57}Fe Mössbauer Spectroscopy

83G1   M. Grimsditch, M.R. Khan, A. Kueny and I.K. Schuller, Phys. Rev.
        Letters, 51(1983)498, Collective Behavior of Magnons in Superlattices

83G2   E.M. Gyorgy, D.B. McWhan, J.F. Dillon Jr., L.R. Walker, J.V. Waszczak,
        D.P. Musser and R.H. Willens, J. Magn. & Magn. Mater., 31-34(1983)915-6,
        Structural and Magnetic Properties of Compositionally Modulated Cu-Ni
        Films

83H1   G.E. Henein and J.E. Hilliard, J. Appl. Phys., 54(1983) 728-37, Elastic
        Modulus in Composition-Modulated Silver-Palladium and Copper-Gold Foils

83H2T   N. Hamada, K. Terakura and A. Yanase, J. Magn. & Magn. Mater., 35(1983)
        7-8, Calculation of Magnetic Moment Distributions in Multilayered Films

83I1   M. Ikebe, N.S. Kazama, Y. Muto and H. Fujimori, IEEE Trans., Mag-19
        (1983)204-7, Mo Base Superconducting Materials Prepared by Multi-Target
        Reactive Sputtering

83J1   M. Jalochowski and P. Mikolajczak, J. Phys., F13(1983) 1973-9,
        The Growth and the X-Ray Diffraction Spectra of the Pb/Ag Superlattice

83J2   M. Jalochowski, Thin Solid Films, 101(1983)285-9, Layered Pb/Ag Films

83K1   N.S. Kazama and H. Fujimori, J. Magn. & Magn. Mater. 35(1983)86-88,
        Diffusion and Magnetic Properties of Compositionally Modulated Films

83K2   M.R. Khan, C.S.L. Chun, G.P. Felcher, M. Grimsditch , A. Kueny,
       C.M. Falco and I.K. Schuller, Phys. Rev. B27(1983)7186–93, Structural,
       Elastic, and Transport Anormalies in Molybdenum/Nickel Superlattices

83M1   D.B. McWhan, M. Gurvitch, J.M. Rowell and L.R. Walker, J. Appl. Phys.,
       54(1983)3886–91, Structure and Coherence of NbAl Multilayer Films

83M2T M. Menon and G.B. Arnold, Phys. Rev. B27(1983)5508–18, Model
       Calculations of Local Electron and Phonon Densities of States in
       Bimetallic Superlattices

83M3T R. Marmoret and J.M. André, Appl. Opt., 22(1983)17–9, Bragg Reflectivity
       of Layered Synthetic Microstructures in the X-Ray Anomalous Scattering
       Regions

83S1R I.K. Schuller and C.M. Falco, Synthetic Metals, 5(1983)205–16,
       Superconducting Properties of Metallic Heterostructures

83S2   I.K. Schuller and A. Rahman, Phys. Rev. Lett., 50(1983)1377–80,
       Elastic-Constant Anomalies in Metallic Superlattices: A Molecular-
       Dynamics Study

83S3T V.F. Sears, Acta Cryst., A39(1983)601–8, Theory of Multilayer Neutron
       Monochromators

83S4   T. Shinjo, N. Hosoito, K. Kawaguchi, T. Takada, Y. Endoh, Y. Ajiro and
       J.M. Friedt, J. Phys. Soc. Jpn., 52(1983)3154–62, Interface Magnetism
       of Fe–Sb Multilayered Films with Artificial Superstructure from ^{57}Fe
       and ^{121}Sb Mössbauer Spectroscopy, Neutron Diffraction and FMR
       Experiments

83S5   T. Shinjo, N. Hosoito and T. Takada, J. Magn. & Magn. Mater., 31–34
       (1983) 879–80, Magnetism of Fe Interfaces Studied by Mössbauer
       Spectroscopy

83S6   J.W. Smits, H.A. Algra, U. Enz and R.P. van Stapele, J. Magn. & Magn.
       Mater., 35(1983)89–92, Ion Beam Sputter Deposition of Layered Magnetic
       Thin Films

83T1   T. Tsakalakos and J.E. Hilliard, J. Appl. Phys., 54(1983)734, Elastic
       Modulus in Composition-Modulated Copper-Nickel Foils

83W1   R. Walmsley, J. Thompson, D. Friedman, R.M. White and T.H. Geballe,
       IEEE Trans., Mag-19(1983)1992, Compositionally Modulated Co-Cr
       Films - A New Co Phase

83Y1   Q.S. Yang, C.M. Falco and I.K. Schuller, Phys. Rev. B27(1983)3867,
       Tunneling Studies of a Metallic Superlattice

84B1   P.R. Broussard, D. Mael and T.H. Geballe, Phys. Rev. B30(1984)4055–6,
       Specific Heat of Niobium-Zirconium Multilayers

84B2   I. Banerjee and I.K. Schuller, J. Low Temp., 54(1984)501–18, Transition
       Temperatures and Critical Fields of Nb/Cu Superlattices

84C1   C.S.L. Chun, G. Zheng, J.L. Vicent and I.K. Schuller, Phys. Rev. B29
       (1984)4915–20, Dimensional Crossover in Superlattice Superconductors

84C2   T. Claeson, J.B. Boyce, W.P. Lowe and T.H. Geballe, Phys. Rev. B29(1984)
       4969–75, NbZr Multilayers. II. Extended X-Ray-Absorption
       Fine-Structure Study

84D1T B.W. Dodson, Phys. Rev. B30(1984)3545–6, Stability of Registry in Strained–Layer Superlattice Interfaces

84E1 Y. Endoh, K. Kawaguchi, N. Hosoito, T. Shinjo, T. Takada, Y. Fujii and T. Ohnishi, J. Phys Soc. Jpn., 53(1984)3481–7, Structure Modulation of Fe–V Artificial Superstructure Films

84F1 C.M. Falco, J. Physique, C5(1984)499–507, Structural and Electronic Properties of Artificial Metallic Superlattices

84F2 C.M. Falco, J. Appl. Phys., 56(1984)1218–9, Figures of Merit for Sputtered Superlattices

84H1 G.E. Henein and J.E. Hilliard, J. Appl. Phys., 55(1984) 2895–900, Interdiffusivities in Silver–Palladium Composition–Modulated Foils

84H2 N. Hosoito, K, Kawaguchi, T. Shinjo, T. Takada and Y. Endoh, J. Phys. Soc. Jpn., 53(1984)2659–67, Magnetic Properties of Fe–V Multilayered Films with Artificial Superstructures

84J1 M. Jalochowski, Phys. Stat. Sol., a82(1984)497–502, Electron Tunneling in Pb/Ag Ultrathin Layered Structures

84J2 B.Y. Jin, H.K. Wong, G.K. Wong, J.E. Hilliard and J.B. Ketterson, J. Appl. Phys., 55(1984)920–25, Preparation and Structural Analysis of SnTe/Sb Composition Modulated Structures

84K1 N.S. Kazama, H. Fujimori, I. Yuito and H. Kronmüller, IEEE Trans., Mag–20(1984)1296–8, Magnetic Properties in Compositionally Modulated Amorphous Fe(C)/Si Alloys

84K2 A. Kueny, M.R. Khan, I.K. Schuller and M. Grimsditch, Phys. Rev. B29 (1984)2879–83, Magnons in Superlattices: A Light Scattering Study

84K3 R. Krishnan and W. Jantz, Solid State Commun., 50(1984)533–5, Ferromagnetic Resonance Studies in Compositionally Modulated Cobalt–Niobium Films

84K4 N. Kumasaka, N. Saito, Y. Shiroishi, K. Shiiki, H. Fujiwara and M. Kudo, J. Appl. Phys., 55(1984)2238–40, Magnetic Properties of Multilayered Fe–Si Films

84K5 R. Krishnan, W. Jantz, W. Wettling and G. Rupp, IEEE Trans., Mag–20 (1984)1264.

84L1 W.P. Lowe and T.H. Geballe, Phys. Rev. B, 29(1984)4961–8, NbZr Multilayers. I. Structure and Superconductivity

84L2T P. Lambin and F. Herman, Phys. Rev. B30(1984)6903–10, Electronic and Magnetic Structure of Idealized Metallic Multilayers: Ni_3Fe–FeMn System

84R1T R. Ramirez, A. Rahman and I.K. Schuller, Phys. Rev. B30(1984)6208–10, Epitaxiy and Superlattice Growth

84R2 R.R.Ruf and R.J.Gambino, J. Appl. Phys., 55(1984)2628–30, Iron–Iron Oxide Layer Films

84S1T Y. Sasajima, M. Imafuku, R. Yamamoto and M. Doyama, J. Phys., F14(1984) L167–72, Computer Simulations of the Dynamical Properties of the Metallic Superlattices, Au/Ni

84S2 I.K. Schuller and M. Grimsditch, J. Appl. Phys., 55(1984)2491-3,
      Magnetic Properties of Mo/Ni superlattices

84S3 T. Shinjo, N. Hosoito, K. Kawaguchi, T. Takada and Y. Endoh,
      J. Physique, C5(1984)361-5, Magnetism of Iron Interface in Contact with
      Vanadium

84S4 T. Shinjo, K. Kawaguchi, R. Yamamoto, N. Hosoito and T. Takada,
      Chem. Lett., (1984)59-62, Synthesis of Fe-Mg Multilayered Films with
      Artificial Superstructures

84S5 T. Shinjo, K. Kawaguchi, R. Yamamoto, N. Hosoito and T. Takada, Solid
      State Commun., 52(1984)257-60, Mössbauer Study of Fe-Mg Multilayered
      Films with Artificial Superstructures

84S6 M.B. Stearns, J. Appl. Phys., 55(1984)1729-31, Mechanism for Enhanced
      Unidirectional Spin-Glass Behavior in Layered Mn-Ni/Co Strucutures

84T1 M.H. Tanielian, J.R. Willhite and D. Niarchos, J. Appl. Phys., 56(1984)
      417-20, Transport Properties of Multilayered Cr/SiO$_x$ Thin Films

84T2 K. Takanashi, H. Yasuoka, K. Takahashi, N. Hosoito, T. Shinjo and
      T. Takada, J. Phys. Soc. Jpn., 53(1984)2445-8, ^{59}Co NMR Observation for
      Compound Formation at Interface of Co-Sb Multilayered Film

84T3 T. Tsakalakos and J.E. Hilliard, J. Appl. Phys., 55(1984)2885-94,
      Effect of Long-Range Interaction on Diffusion in Copper-Nickel
      Composition-Modulated Alloys

84T4 K. Takanashi, H. Yasuoka, K. Kawaguchi, N. Hosoito and T. Shinjo,
      J. Phys. Soc. Jpn., 53(1984)4315-21, Microscopic Magnetic Properties of
      Fe-V Metallic Superlattice Investigated from ^{51}V NMR

84U1 C. Uher, R. Clarke, G.-G. Zheng and I.K. Schuller, Phys. Rev. B30(1984)
      453-5, Interplay of Superconductivity, Magnetism, and Localization in
      Mo/Ni Superlattices

84W1 H.K. Wong, H.Q. Yang, B.Y. Jin, Y.H. Shen W.Z. Cao, J.B. Ketterson and
      J.E. Hilliard, J. Appl. Phys., 55(1984)2494-6, V/Fe Composition-
      Modulated Structures

84Y1T A.L. Yeyati, N.V. Cohan and M. Weissmann, Phys. Rev. B31(1984)873-8,
      Electronic Densities of States of Bimetallic Superlattices with
      Interfacial Diffusion

85A1 R. Akihama and Y. Okamoto, Solid State Commun., 53(1985)655-9,
      Superconductivity in Au(10A)/Ge(13A) Alternating Ultra-Thin Layered
      Films

85B1 D. Baral, J. B. Ketterson and J.E. Hilliard, J. Appl. Phys., 57(1985)
      1076-83, Mechanical Properties of Composition Modulated Cu-Ni Foils

85C1 R. Clarke, D. Morelli, C. Uher, H. Homma and I.K. Schuller,
      Superlattices and Microstructures, 1(1985)125-9, Electronic Transport in
      Mo/Ni Superlattices

85C2 P.F. Carcia, A.D. Meinhaldt and A. Suna, Appl. Phys. Letters, 47(1985)
      178-80, Perpendicular Magnetic Anisotropy in Pd/Co Thin Film Layered
      Structures

85C3 P.F. Carcia, A. Suna, D.G. Onn and R. van Antwerp, Superlattices and

Microstructures, 1(1985)101-9, Structural, Magnetic, and Electrical
Properties of Thin Film Pd/Co Layered Structures

85D1T B. Djafari-Rouhani, L. Dobrzynski and P. Masri, Phys. Rev. B31(1985)
7739-48, Theory of Surface Electronic States in Metallic Superlattices

85D2 H.J.G. Draaisma, H.M. van Noort and F.J.A. den Broeder, Thin Solid
Films, 126(1985)117-21, Magnetic, Microstructural and Mössbauer
Studies of Cu/Fe Composition-Modulated Thin Films

85D3T R. Dimmich, J. Phys. F: Met. Phys., 15(1985)2477-87, Electronic
Transport Properties of Metallic Multi-Layer Films

85G1 L.H. Greene, W.L. Feldmann, J.M. Rowell, B. Batlogg, E.M. Gyorgy,
W.P. Lowe and D.B. McWhan, Superlattices and Microstructures, 1(1985)
407-15, Structural, Magnetic and Superconducting Properties of Rare
Earth/superconductor Multilayer Films

85H1T F. Herman, P. Lambin and O. Jepsen, Phys. Rev. B31(1985)4394-402,
Electronic and Magnetic Structure of Ultrathin Cobalt-Chromium
Superlattices

85H2T F. Herman, P. Lambin and O. Jepsen, J. Appl. Phys., 57(1985)3654-6,
Electronic and Magnetic Structure of Ultrathin Cobalt-Chromium
Multilayers

85H3 H. Homma, C.S.L. Chun, G.-G. Zheng and I.K. schuller, Phsica, 135B(1985)
173-7, Dimensional and Magnetic Efects in Superconducting Superlattices

85J1 N.K. Jaggi, L.H. Schwartz, H.K. Wong and J.B. Ketterson, J. Magn. &
Magn. Mater., 49(1985)1-14, Mössbauer Spectroscopy Study of Composition
Modulated   110   Fe-V Films

85J2 N.K. Jaggi and L.H. Schwartz, J. Phys. Soc. Jpn., 54(1985)1652-3,
Magnetism in Composition Modulated Superlattices of Fe-V: A Comment

85J3 B.Y. Jin, Y.H. Shen, H.Q. Yang, H.K. wong, J.E. Hilliard, J.B. Ketterson
and I.K. Schuller, J. Appl. Phys., 57(1985)2543, Superconducting
Properties of Layered $Nb_{0.53}Ti_{0.47}$/Ge Structures Perpared by Dc
Sputtering

85K1 N.S. Kazama, H. Fujimori, I. Yuito and H. Kronmüller, Sci. Rep. Ritu.,
A-32(1985)141-53, Preparation and Magnetic Properties of Amorphous
Superlattice

85K2 R. Krishnan, J. Magn. & Magn. Mater., 50(1985)189-92, FMR Studies in
Compositionally Modulated Co-Nb and Co Films

85K3 J. Kwo, E.M. Gyorgy, D.B. McWhan, M. Hong, F.J. DiSalvo, C. Vettier and
J.E. Bower, Phys. Rev. Lett., 55(1985)1402-5, Magnetic and Structural
Properties of Single-Crystal Rare-Earth Gd-Y Superlattices

85K4 M.G. Karkut, D. Ariosa, J.M. Triscone and O. Fischer, Phys. Rev. B32
(1985)4800-3, Epitaxial Growth and Superconducting-Transition-
Temperature Anomalies of Mo/V superlattices

85K5 M.G. Karkut, J.M. Triscone, D. Ariosa and Ø. Fischer, Physica, 135B
(1985) 182-4, Superconducting T_c Anomalies in Mo/V Superlattices

85L1 W.P. Lowe, E.M. Gyorgy, D.B. McWhan, L.H. Greene, W.L. Feldman and
J.M. Rowell, J. Appl. Phys., 58(1985)1615-8, Magnetic and Structural

Properties of Tm_nLu_m Multilayer Films

85M1   C.F. Majkrzak, J.D. Axe and P. Böni, J. Appl. Phys., 57(1985)3657-9,
       Magnetic Structure of Multiple bilayers of Thin Films of Fe and Ge

85M2   P.F. Miceli, H. Zabel and J.E. Cunningham, Phys. Rev. Lett., 54(1985)
       917-9, Hydrogen-Induces Strain Modulation in Nb-Ta Superlattices

85M3   T. Morishita, Y. Togami and K. Tsushima, J. Phys. Soc. Jpn., 54(1985)
       37-40, Magnetism and Structure of Compositionally Modulated Fe-Gd Thin
       Films

85M4   C.F. Majkrzak and L. Passell, Acta Cryst., A41(1985)41-8, Multilayer
       Thin Films as Polarizing Monochromators for Neutron

85M5T K. Mika and P. Grünberg, Phys. Rev. B31(1985)4465-71, Dipolar
       Spin-Wave Modes of a Ferromagntic Multilayer with Alternating Directions
       of Magnetization

85M6   S. Moehlecke, C.F. Majkrzak and M. Strongin, Phys. Rev. B31(1985)6804-6,
       Enhanced Hydrogen Solubility in Niobium Films

85N1   H.M. van Noort ,F.J.A. den Broeder and H.J.G. Draaisma, J. Magn. & Magn.
       Mater., 51(1985)273-9, Mössbauer Study of Cu-Fe Composition-Modulated
       Thin Films

85P1   M.J. Pechan, M.B. Salamon and I.K. Schuller, J. Appl. Phys., 57(1985)
       3678-89, Ferromagnetic Resonance in a Ni-Mo Superlattice

85S1   H. Sakakima, R. Krishnan and M. Tessier, J. Appl. Phys., 57(1985)3651-3,
       Magnetic Properties of Compositionally Modulated Co/Mn Thin films

85S2   T. Shinjo, N. Hosoito and Y. Endoh, J. Phys. Soc. Jpn., 54(1985)1654-5,
       Reply to the Comment by Jaggi and Schwartz

85S3   R.P. van Stapele, F.J.A.M. Greidanus and J.W. Smits, J. Appl. Phys.,
       57(1985)1282-90, The Spin-Wave Spectrum of Layered Magnetic Thin Films

85S4   T. Shinjo, K. Kawaguchi, R. Yamamoto, N. Hosoito and T. Takada,
       Thin Solid Films, 125(1985)273-6, Preparation and Characterization of
       Fe/Mg Superlattice Films

85T1   K. Takei and Y. Maeda, Jpn. J. Appl. Phys., 24(1985)118-9, Preparation
       of Multi-Layered Tungsten-Carbon Films by Ion-Beam Sputterin

85T2   M. Takahashi, S. Ishio and Y. Notohara, Proc. 2nd Intern. Conf. Physics
       of Magnetic Materials, Jadwisin, 1984, World Sci. Pub. Singapore
       (1985)13, Co-Cr and Co-Ag Compositionally Modulated Films

85T3T M. Tachiki and S. Takahashi, Physica, 135B(1985)178-80, Theory of the
       Upper Critical Field of Superconducting Superlattices

85U1   S. Umemura, H. Tajika, E. Kita and A. Tasaki, IEEE Trans. Mag., Mag-21
       (1985)1942-4, Magnetic Properties of Alternately Evaporated Fe-Gd Films

85V1T D.M. Vardanyan, H.M. Manoukyan and H.M. Petrosyan, Acta Cryst.,
       A41(1985)212-7, The Dynamic Theory of X-Ray Diffraction by the One-
       Dimensional Ideal Superlattice. I. Diffraction by the Arbitrary
       Superlattice

85V2T D.M. Vardanyan, H.M. Manaukyan and H.M. Petrosyan, Acta Cryst.,

A41(1985)218-22, The Dynamic Theory of X-Ray Diffraction by the One-Dimensional Ideal superlattice. II. Calculation of Structure Factors for Some Superlattice Models

85V3 S.P. Vernon, B.N. Halawith and M.B. Stearns, J. Appl. Phys., 57(1985) 3441-3, Spin-Glass and Ferromagnetic Behavior of Ni-Mn Multilayered Films

85W1 H.K. Wong, H.Q. Yang, J.E. Hilliard and J.B. Ketterson, J. Appl. Phys., 57(1985)3660-77, Magnetic Properties of V/Fe Superlattices

85W2 D.J. Webb, R.G. Walmsley, K. Parvin, P.H. Dickinson, T.H. Geballe and R.M. White, Phys. Rev. B32(1985)4667-75, Sequential Deposition and Metastable States in Rare-earth/Co Films

85Y1 H. Yamamoto, M. Ikeda and M. Tanaka, Jpn. J. Appl. Phys., 24(1985) L314-6, Giant Resistivity Anomaly in A15 Nb_3(Ge, Si) Superconductive Films with Compositionally Modulated Superstructure

85Y2 M. Yudkowsky, W.P. Halperin and I.K. Schuller, Phys. Rev. B31(1985) 1637-9, Nuclear-Magnetic-Resonance Study of Electronic Structure in the Copper-Niobium Superlattice

85Z1 C.B. Zimm, M.B. Stearns and P.R. Roach, J. Magn. & Magn. Mater., 50(1985)223-8, Magnetization of Layered Mn-Ni and Mn-Co Thin Films

86A1 Y. Asada and K. Ogawa, Solid State Commun., 60(1986)161-4, Superconductivity of Mo/Sb Multilayered Films

86A2T P. Apell and C. Holmberg, Superlattices and Microstructures, 2(1986) 297-301, Bulk and Surface Collective Modes in Metal Superlattices

86B1T K.R. Biagi, J.R. Clem and V.G. Kogan, Phys. Rev. B33(1986)3100-1, Perpendicular Upper Critical Field of Thick Proximity-Coupled Multilayers

86B2 P. Baumgart, B. Hillebrands, R. Mock, G. Gutherodt, A. Boufelfel and C.M. Falco, Phys. Rev. B34(1986)9004-7, Localized Phonon Modes in Fe-Pd Multilayer Structures

86C1 J.W. Cable, M.R. Khan, G.P. Felcher and I.K. Schuller, Phys. Rev. B34 (1986)1643-9, Macromagnetism and Micromagnetism in Ni-Mo Metallic Superlattices

86C2 R. Clarke, F. Lamelas, C. Uher, C.P. Flynn and J.E. Cunningham, Phys. Rev. B34(1986)2022-5, Stacking Structure and Superconductivity in Ruthenium-Iridium Bicrystal Superlattices

86D1 K.Le Dang, P. Veillet, H. Sakajima and R. Krishnan, J. Phys. F: Met. Phys., 16(1986)93-7, NMR Studies of Compositionally Modulated Co/Mn Thin Films

86D2 R. Danner, R.P. Huebener, C.S.L. Chun, M. Grimsditch and I.K. Schuller, Phys. Rev. B33(1986)3696-701, Surface Acoustic Waves in Ni/V Superlattices

86E1T M.E. Elzain, D.E. Ellis and D. Guenzburger, Phys. Rev. B34(1986)1430-41, Electronic Structure and Magnetic and Hyperfine Properties of Fe/V Sandwiches and Interfaces

86F1 Y. Fujii, T. Ohnishi, T. Ishihara, Y. Yamada, K. Kawaguchi, N. Nakayama and T. Shinjo, J. Phys. Soc. Jpn., 55(1986)251-62, Structural Aspects of Fe-Mg Artificial Superstructure Films Studied by X-Ray Diffraction

86H1T   L.L. Hinchey and D.L. Mills, Phys. Rev. B34(1986)1689-99, Magnetic Properties of Ferromagnet-Antiferromagnet Superlattices Structures with Mixed-Spin Antiferromagnetic Sheets

86H2   T. Hatano, K. Nakamura, Y. Asada, K. Ogawa and A. Oguchi, Jpn. J. Appl. Phys., 25(1986)912-3, Characterization of Cu/Ti Multi-Layered Structure by Improved Sputter Crater Method

86H3T   L.L. Hinchey and D.L. Mills, Phys. Rev. B33(1986)3329-43, Magnetic Properties of Superlattices Formed from Ferromagnetic and Antiferromagnetic Materials

86H4   B. Hillebrands, P. Baumgart, R. Mock, G. Güntherodt, A. Boufelfel and C.M. Falco, Phys. Rev. B34(1986)9000-3, Collective Spin Waves in Fe-Pd and Fe-W Multilayer Structure

86K1   K. Kanoda, H. Mazaki, T. Yamada, N. Hosoito and T. Shinjo, Phys. Rev. B33(1986)2052-5, Dimensional Crossover and Commensurability Effect in V/Ag Superconducting Multilayers

86K2   N.S. Kazama and H. Fujimori, J. Magn. & Magn. Mater., 54-7(1986)793-4, Magnetic and Diffusional Properties of Compositionally Modulated Amorphous CoNb/CoTi Alloys

86K3   J. Kwo, E.M. Gyorgy, F.J. DiSalvo, M. Hong, Y. Yafet and D.B. McWhan, J. Magn. & Magn. Mater., 54-7(1986)771-2, Magnetic Properties of Single Crystal Rare-Earth Gd-Y Superlattices

86K4   J. Kwo, D.B. McWhan, M. Hong, E.M. Gyorgy and F.J. DiSalvo, Mat. Res. Soc. symp. Proc., 56(1986)211-6, Structural and Magnetic Properties of Single Crystal Rare earth Gd-Y Superlattices

86K5   K. Kawaguchi, R. Yamamoto, N. Hosoito, T. Shinjo and T. Takada, J. Phys. Soc. Jpn., 55(1986)2375-83, Magnetic Properties of Fe-Mg Artificial Superstructure Films

86K6   J. Kwo, M. Hong and S. Nakahara, Appl. Phys. Lett., 49(1986)319-21, Growth of Rare-earth Single Crystals by Molecular Beam Epitaxy: The Epitaxial Relationship between hcp Rare Earth and bcc Niobium

86K7   T. Katayama, H. Awano and Y. Nishihara, J. Phys. Soc. Jpn., 55(1986) 2539-42, Wavelength Dependence of Magneto-Optical Kerr Rotation in Co/Cu, Fe/Cu, Co/Au and Fe/Au Compositionally Modulated Multilayered Films

86K8   M.G. Karkut, J.-M. Triscone, D. Ariosa and Ø. Fischer, Phys. Rev. B34 (1986)4390-3, Quasiperiodic Metallic Multilayers: Growth and Superconductivity

86K9   R. Krishnan and M. Tessier, Solid State Commun., 60(1986)637-9, Magnetization and FMR Studies in Multilayer Ni-Ag Films

86L1   R.H.M. van de Leur, A.J.G. Schellingerhout, J.E. Mooij and F. Tuinstra, Solid State Commun., 60(1986)633-5, Growth and Characterization of A Sinusoidally Modulated Nb/V Superlattice

86M1   D.B. McWhan, C. Vettier, E.M. Gyorgy, J. Kwo, B. Buntschuh and B. Batterman, J. Magn. & Magn. Mater., 54-7(1986)775-6, Magnetic X-Ray

Scattering from Superlattices

86M2   T. Morishita, Y. Togami and K. Tsushima, J. Magn. & Magn. Mater.,
       54-7(1986)789-90, Magnetic Properties of Compositionally Modulated
       Gd-Fe and Y-Fe Films

86M3   C.F. Majkrzak, J.W. Cable, J. Kwo, M. Hong, D.B. McWhan, Y. Yafet,
       J.V. Waszczak and C. Vettier, Phys. Rev. Lett., 56(1986)2700-3,
       Observation of a Magnetic Antiphase Domain Structure with Long-Range
       Order in a Synthetic Gd-Y Superlattice

86M4R C.F. Majkrzak, Physica, 136B(1986)69-74, Neutron Diffraction Studies of
       Thin Film Multilayer Structures

86N1   N. Nakayama, K. Takahashi, T. Shinjo, T. Takada and H. Ichinose,
       Jpn. J. Appl. Phys., 25(1986)552-7, A Structural Study of Co-Sb
       Multilayered Film by X-Ray Diffraction

86N2   D. Niarchos, Solid State Commun., 59(1986)81-4, Hall, Electrical
       Resistivity and Magnetoresistivity Measurements in Mo/Si_3N_4 and Ti/Si_3N_4
       Multilayered Thin Films

86O1T T. Oguchi and A.J. Freeman, J. Magn. & Magn. Mater., 54-57(1986)797-8,
       Interface Magnetism of Metallic Superlattices: CrAu

86S1   H. Sakakima, R. Krishnan, M. Tessier, K.L. Dang and P. Veillet, J. Magn.
       & Magn. Mater., 54-7(1986)785-6, Anomalous Temperature Dependence of
       Magnetization in Multilayered Co/Mn Thin Films

86S2   M.B. Salamon, S. Sinha, J.J. Rhyne, J.E. Cunningham, R.W. Erwin,
       J. Borchers and C.P. Flynn, Phys. Rev. Lett., 56(1986)259-62, Long-Range
       Incommensurate Magnetic Order in a Dy-Y Multilayer

86S3   N. Sato, J. Appl. Phys., 59(1986)2514-20, Magnetic Properties of
       Amorphous of Tb-Fe Thin Films with an Artificially Layered Structure

86S4   C. Sellers, Y. Shiroishi, N.K. Jaggi, J.B. Ketterson and J.E. Hilliard,
       J. Magn. & Magn. Mater., 54-7(1986)787-8, Magnetic Properties of
       Compositionally Modulated Fe/Cr Thin Films

86S5T J. Simonin, Phys. Rev. B33(1986)1700-5, Theory for Superconductivity
       in Metallic Multilayer Compounds

86S6   S. Sinha, J. Cunningham, R. Du, M.B. Salamon and C.P. Flynn, J. Magn. &
       Magn., 54-7(1986)773-4, Magnetic Properties of Dy-Y Superlattice
       Structures

86S7   T. Shinjo, Hyperfine Interactions, 27(1986)193-202, Metallic
       Superlattices

86S8   T. Shinjo, N. Hosoito, K. Kawaguchi, N. Nakayama, T. Takada and
       Y. Endoh, J. Magn. & Magn. Mater., 54-7(1986)737-42, Magnetic Properties
       of Artificial Metallic Superlattices

86S9   M.B. Stearns, C.H. Lee and S.P. Vernon, J. Magn. & Magn. Mater., 54-57
       (1986)791-2, Magnetic and Structural Properties of Co-Cr Multilayered
       Structures

86S10 T. Shinjo, N. Nakayama, I. Moritani and Y. Endoh, J. Phys. Soc. Jpn.,
       55(1986)2512-4, Monolayer of Ferromagnetic MnSb

86T1T S. Takahashi and M. Tachiki, Phys. Rev. B33(1986)4620-31, Theory of the Upper Critical Field of Superconducting Superlattices

86T2 K. Takanashi, H. Yasuoka and T. Shinjo, J. Magn. & Magn. Mater., 54-7(1986)783-4, NMR Investigations for New-Type Alloy Formation at the Interface in Artificial Superstructure Film

86T3 K. Takanashi, H. Yasuoka, N. Nakayama, T. Katamoto and T. Shinjo, J. Phys. Soc. Jpn., 55(1986)2357-63, Interface Magnetism in Fe/Mn Artificial Metallic Superlattice Investigated by ^{55}Mn NMR

86U1 C. Uher, W.J. Watson, J.L. Cohn and I.K. Schuller, Mat. Res. Soc. Symp. Proc., 58(1986)455-60, Upper Critical Firld of Mo-Ni Heterostructures

86U2 S. Umemura, H. Tajika, E. Kita and A. Tasaki, Advances in Ceramics 16(1986)621-5, Magnetic Properties of Alternately Evaporated Fe-Sm Film

86U3 C. Uher, J.L. Cohn and I.K. Schuller, Phys. Rev. B34(1986)4906-9, Upper Critical Field in Anisotropic Superconductors

86V1 C. Vettier, D.B. McWhan, E.M. Gyorgy, J. Kwo, B.M. Buntschuh and B.M. Batterman, Phys. Rev. Lett., 56(1986)757-60, Magnetic X-Ray-Scattering Study of Interfacial Magnetism in a Gd-Y Superlattice

86V2T K. Vayhinger and H. Kronmüller, J. Magn. & Magn. Mater., 62(1986)159-68, Propagating Spin Waves in Ferromagnetic Multilayers

86Z1 G.G. Zeng, H. Xia, Y.Z. Wei and J.Q. Zeng, Superlattices and Microstructures, 2(1986)483-9, Structure Analysis of Mo/Ni Superlattice

Subject Index